Circuits and Systems for Wearable Technologies

IEEE UKCAS 2019

EDITORS

Sara Ghoreishizadeh
University College London

Kylie de Jager
University College London

Tutorials in Circuits and Systems

For a list of other books in this series, visit www.riverpublishers.com

Series Editors

Peter (Yong) Lian

President IEEE Circuits
and Systems Society
York University, Canada

Franco Maloberti

Past President IEEE Circuits and
Systems Society
University of Pavia, Italy

LONDON AND NEW YORK

Published 2019 by River Publishers
River Publishers
Alsbjergvej 10, 9260 Gistrup, Denmark
www.riverpublishers.com

Distributed exclusively by Routledge
4 Park Square, Milton Park, Abingdon, Oxon OX14 4RN
605 Third Avenue, New York, NY 10017, USA

Circuits and Systems for Wearable Technologies IEEE UKCAS 2019 / by Sara
Ghoreishizadeh, Kylie de Jager, Peter (Yong) Lian, Franco Maloberti.

Routledge is an imprint of the Taylor & Francis Group, an informa
business

ISBN 9788770221320 (print)

While every effort is made to provide dependable information, the publisher, authors, and
editors cannot be held responsible for any errors or omissions.

Table of contents

Introduction

ecent years, wearable technologies have gained popularity, finding application in numerous lds including fitness, healthcare, security and navigation. As a result, increased interest ircuit and system techniques relevant to wearables has led to advances in lab-on-chip nologies, sensors, real-time data analysis and energy considerations; to name a few. Within healthcare field these technologies have been applied to both wearable and implantable nedical systems although various technical challenges still exist that must be overcome for e systems to come to fruition.

he United Kingdom Circuits and Systems (UKCAS) Workshop provides a platform for archers and academic participants from UK and Ireland universities to come together and ange knowledge, innovative solutions and the latest developments around the challenges in emerging wearable technologies. In line with the mission of our parent body, IEEE CASS uits and Systems Society), this includes ideas that contribute toward the "advancement of ry, analysis, design, tools, and implementation of circuits and systems", within the realm of thcare and biomedical applications.

nis book proceeds the 2ⁿᵈ UKCAS Workshop (https://attend.ieee.org/ukcas/). It is a collection ne presentations given at the workshop by leading UK and Ireland academics in the field. ɔugh the applications discussed are diverse, the research presented has several common it and systems related themes.

napter 1 discusses microchip technology for the rapid diagnosis of infectious diseases. **CMOS** ·d lab-on-chip systems use **ISFETs** (Ion Sensitive Field Effect Transistors) to realise **multi-ʒor** platforms in the area of DNA detection to advance point-of-care diagnostics.

on-invasive, wearable, **bioelectronic sensors**, used to automate and personalise treatments ndividuals, are presented in Chapter 2. The technologies considered move **beyond CMOS** vestigating the confluence of several emerging technologies: **flexible electronics** (discrete personalised wearables); **low power consumption signal processing** (improving battery- and, **real-time data analysis** (feedback and actuation).

 Chapter 3 the use of **ISFETs** for realising micro- and nano-scale **bio-sensors** are explored onsidering two applications. Firstly, miniature implantable **sensors** that measure oxygen pH in tumours are shown to assist with optimising radiotherapy delivery. Secondly, the **inuous monitoring** of biomarkers is achieved using an electrochemical contact lens.

napter 4 looks at the non-invasive monitoring of ECG in foetuses and neonates. Novel **circuit ·logies** are presented that remove the need for post processing, an advantageous step rds **real-time data analysis**.

ɪrther **circuit topologies** for the remote sensing of forces in orthopaedic implants that tate wireless power and data transfer are covered in Chapter 5. Several design challenges are discussed: development of implanted and external circuits; factors influencing longevity of implanta devices; and, **calibration** of force transducers.

Wearable and implantable devices that move **beyond-CMOS** technologies are the focus of Chapter 6. efficiency, in terms of size, **performance and power**, of traditional CMOS technology used in neural electrc applications, can be improved on using: spintronics and biocompatible materials; **magnetic sensors**; a **flexible sensors**.

In Chapter 7 nanowires are used to realise **biological sensors** by using dielectrophoresis as a manufactur approach. The nanowires are used on **lab-on-a-chip** and wearable technologies toward diagnostic monitoring applications. The nanowire-based sensors have the potential for **real-time, continuous monitori**

The final two chapters both explore **energy requirements** of wearable tech. Chapter 8 looks at ene driven computing wherein **energy storage** requirements are reduced by including the energy harves sources in the system's design. Chapter 9 considers techniques for harvesting kinetic energy from irreg motion, such as human gait. The mechanical and electrical components of an **energy harvester** are preser as well as transduction mechanisms and system level requirements.

We hope this book proves a useful tool for all those working on biomedical devices, assisting research scientists, and engineers in translating their ideas into a viable product that reaches the end user and real envisioned healthcare advancements.

We are grateful to the authors and speakers of the 2nd UKCAS Workshop. Their willingness to contrit their time and energy, made this book possible, while their commitment to sharing expertise and knowle fosters further advancement of the wearable biomedical device field. We also extend our thanks to t organisations and colleagues, who support their research on a daily basis. The compilation and editing of book was supported by IEEE CASS and River Publishers.

SARA GHOREISHIZADE
KYLIE DE JAGE
NOVEMBER 201

CHAPTER 01

Microchip Technology Enabling Rapid Diagnostics For Infectious Diseases

Pantelis Georgiou

Imperial College London

In the last decade, we have seen a convergence of microelectronics into the world of healthcare providing novel solutions for early detection, diagnosis and therapy of disease. This has been made possible due to the emergence of CMOS technology, allowing fabrication of advanced systems with complete integration of sensors, instrumentation and processing, enabling design of miniaturised medical devices which operate with low-power. This has been specifically beneficial for the application areas of DNA based diagnostics and full genome sequencing, where the implementation of chemical sensors known as Ion-Sensitive Field Effect Transistors (ISFETs) directly in CMOS has enabled the design of large-scale arrays of millions of sensors that can conduct in-parallel detection of DNA. Furthermore, the scaling of CMOS with Moore's law and the integration capability with microfluidics has enabled commercial efforts to make DNA detection affordable and therefore deployable in hospitals and research labs.

In this talk, I present how my lab is advancing the areas of DNA detection and rapid diagnostics through the design of CMOS based Lab-on-Chip systems using ISFETs. I will present methods of design of ISFET sensors and instrumentation in CMOS and showcase the state-of-the-art CMOS systems that are currently being used for genomics and point-of-care diagnostics. I will conclude with the results of our latest fabricated multi-sensor CMOS platform for rapid screening of infectious disease and management of antimicrobial resistance.

1

We've all been there....

2

Microchips – Nano-Technology enabling new directions

A11 Bionic

Superhuman intelligence.

3 — My Lab

Bio-inspired design

b·iap
bio inspired artificial pancreas

Integrated Sensing Systems

Wearable medical devices

---

4 — The need for Rapid Diagnostics

- **Early and accurate diagnosis** is essential for disease management and control.
- It is curable if **treated promptly and accurately!**
- Reduce **drug resistance!**
- Reduce **the length of hospital stays.**
- Limit **nosocomial infections.**
- Essential for **remote diagnosis and outbreak investigations.**

World Health Organization

WHO recommends prompt confirmation using rapid diagnostic test (RDT)

5 Rapid diagnostics of the 15th Century!

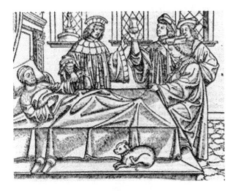

Real time monitoring and diagnosing ?

Uroscopy wheel

6 Diagnostics – Desirable Characteristics

The ASSURED criteria

Affordable (<$10/test)
Sensitive (>95%)
Specific (>95%)
User-friendly (minimal training and easy to read)
Rapid (<2h; single patient-health care contact)
Robust (shelf-life >24 months)
Equipment-free (portable handheld size)
Deliverable to end users

World Health Organization

7 Rural health – The real world

8 Current Diagnostic Techniques

Conventional Methods

Molecular methods (PCR)

Rapid Detection Tests (RDTs)

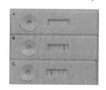

+

Conventional Methods	Molecular methods (PCR)	Rapid Detection Tests (RDTs)
Time (up to few days)	Time (>2h)	Rapid (15-30min)
Expertise	Expertise	No expertise
Lab dependent	High cost	Affordable
User dependent	Lab dependent	Portable
Specie & AMR identification	Specie & AMR identification	Specie & AMR identification
Sensitivity	High sensitivity	Low sensitivity
	High specificity	Qualitative
	Quantitative	

9 Microchips for DNA diagnostics

- **Point-of-Care Suitability**
  - Can produce a small (handheld) device without bulky optics
  - Lab free
- **Real-Time Detection Capability**
  - Instrumentation is integrated directly with sensor
- **Implementation of large arrays**
  - Integrate arrays over 1000 sensors
- **Opportunities**
  - Proven by our group for **reliable DNA detection**
  - Pioneered **Ion-Sensitive Field Effect Transistors**

"Simultaneous DNA amplification and detection using a pH-sensing semiconductor system," Nature methods, 2013.

ISFET Bio-Sensors

10 Enabler: The ion sensitive field effect transistor

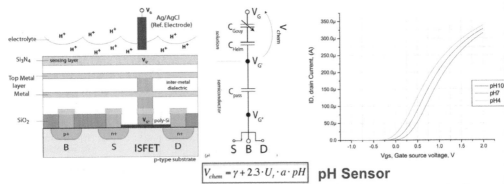

$$V_{chem} = \gamma + 2.3 \cdot U_t \cdot a \cdot pH$$

pH Sensor

Pantelis Georgiou, Christofer Toumazou, ISFET characteristics in CMOS and their application to weak inversion operation, Sensors and Actuators B: Chemical, Volume 143, Issue 1, 4 December 2009

11 The Key - pH based nucleic acid detection

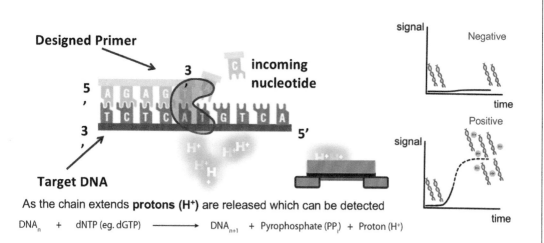

Designed Primer

incoming nucleotide

Target DNA

As the chain extends **protons (H⁺)** are released which can be detected

$$DNA_n + dNTP (eg. dGTP) \longrightarrow DNA_{n+1} + Pyrophosphate (PP_i) + Proton (H^+)$$

12 Advantages of pH based methods

- Chain extension allows detection of binding simply by monitoring the **pH change**.
- This requires **no labeling** of the target molecules to detect their presence.
- This greatly **simplifies** the sample preparation process.
- Termed **Label Free detection** and requires **no optics**.
- Can be detected using bound probes or even **free floating** probes in a solution.
- Can be implemented using sensors fabricated in **CMOS!**
- Incredibly **scalable** to thousands of sensors.

CIRCUITS AND SYSTEMS FOR WEARABLE TECHNOLOGIES – IEEE UKCAS 2019

13 Commercial DNA detection platforms using ISFETs

Next-Generation Sequencing: **massively parallel** and **fully electronic**

Genotyping

Sequencing

[1]

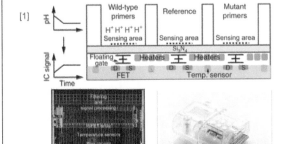

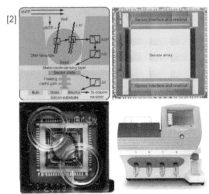

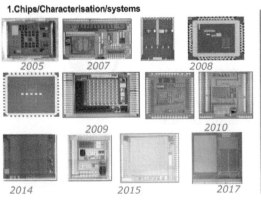

[1] Toumazou et al, Nature Methods, 2013.
[2] Rothberg et al, Nature, 2011.

---

14 ISFET research – a story of over a decade

1. Chips/Characterisation/systems

2005   2007   2008

2009   2010

2014   2015   2017

2. Bonding/ Packaging

3. Surface Deposition/Reference Electrodes

15 ## Multimodal sensing on CMOS

A Novel Pixel architecture combining Chemical, Optical, Magnetic sensing.

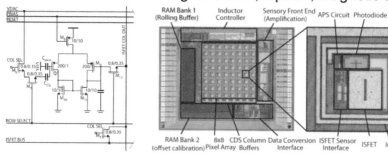

RAM Bank 1 (Rolling Buffer)   Inductor Controller   Sensory Front End (Amplification)   APS Circuit   Photodiode

RAM Bank 2 (offset calibration)   8x8 Pixel Array   CDS Column Buffers   Data Conversion Interface   ISFET Sensor Interface   ISFET   Inductor

, Z. D. Goh, P. Georgiou, T. G. Constandinou, T. Prodromakis and C. Toumazou, "A CMOS-Based ISFET Chemical Imager With Auto-Calibration Capability," in *IEEE Sensors Journal*, vol. 11, no. 12, pp. 3253-3260, Dec. 2011.

16 ## Multimodal sensing – pH, optical, magnetic

uto calibration capability
- Gradient based
- Mismatch compensation
- Drift compensation

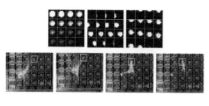

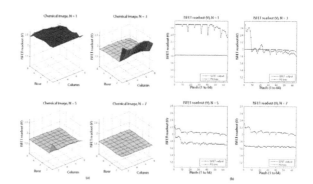

Z. D. Goh, P. Georgiou, T. G. Constandinou, T. Prodromakis and C. Toumazou, "A CMOS-Based ISFET Chemical Imager With Auto-Calibration Capability," in *IEEE Sensors Journal*, vol. 11, no. 12, pp. 3253-3260, Dec. 2011.

17 ISFET array with trapped charge and gain compensation

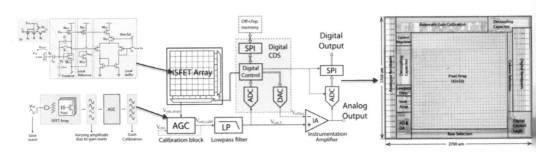

- Y. Hu and P. Georgiou, "An Automatic Gain Control System for ISFET Array Compensation," in *IEEE Transactions on Circuits and Systems I: Regular Papers*, vol. 63, no. 9, pp. 1511-1520, Sept. 2016.
- Y. Hu, N. Moser and P. Georgiou, "A 32 x 32 ISFET Chemical Sensing Array With Integrated Trapped Charge and Gain Compensation," in *IEEE Sensors Journal*, vol. 17, no. 16, pp. 5276-5284, Aug.15, 15 2017.

18 Robust ISFET arrays for ion-imaging

- **You can now for the first time see the chemical reaction!**

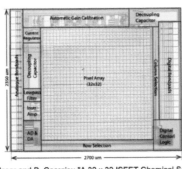
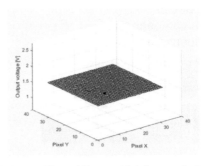

Y. Hu, N. Moser and P. Georgiou,"A 32 x 32 ISFET Chemical Sensing Array With Integrated Trapped Charge and Gain Compensation"in *IEEE Sensor Journal*, vol.17,no.16, pp.5276-5284, Aug.15,15 2017

19 ## Multi-ion imaging – Sodium and Potassium

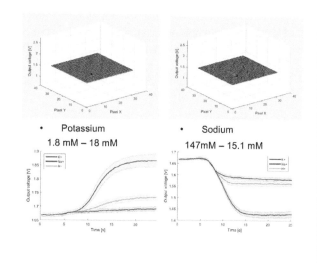

- Potassium
  1.8 mM – 18 mM

- Sodium
  147mM – 15.1 mM

N.Moser, An ion imaging ISFET array for
Potassium and Sodium detection (2016). In
*Proceedings - IEEE International Symposium
on Circuits and Systems* Vol. 2016-July (pp.
2847-2850).

20 ## Challenge - Sample preparation

- **Cell Lysis**- Break Cells open to extract DNA
- **Cleaning and purification** - Remove lipids, proteins, RNA
- **Elution** – extraction of the DNA sample from the solution

- This will give you the DNA sample you would then want to investigate
- **If copies of DNA is low it may be undetectable!**

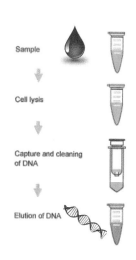

Sample

Cell lysis

Capture and cleaning
of DNA

Elution of DNA

21 Boosting the DNA signal

- Signals can be amplified by increasing the number of copies of DNA.
- This is achieved by **Polymerase Chain Reaction (PCR)** which increases the number of copies of DNA by replication.
- Requires cycling of temperature chamber from 60-90 °C

Polymerase chain reaction - PCR

Agilent qPCR

Source: https://en.wikipedia.org/wiki/Polymerase_chain_reaction

22 DNA Detection with amplification in CMOS

a

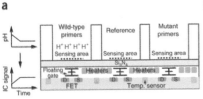

Features
- Detect Hydrogen ions using Ion-Sensitive Field Effect Transistors.
- Can use heating to amplify DNA through PCR (polymerase chain reaction).
- Use a reference chamber to do differential measurement and cancel out chemical drift.

b

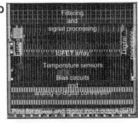

C.Toumazou, P Georgiou et al.,"Simultaneous DNA amplification and detection using a pH-sensing semiconductor system," Nature methods, 20

23 Keeping with the trends - Next generation ISFET arrays

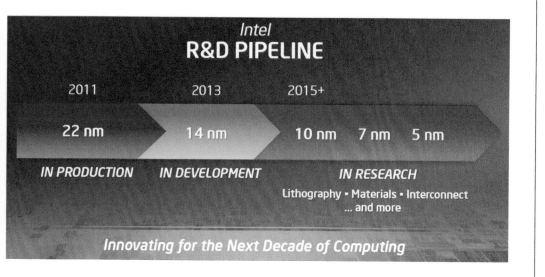

24 pH-Time: Scaling to deep-submicron technology

- Closed loop compensation of trapped charge and drift

On-chip algorithm to find the source voltage V_s that compensates for the offset in each pixel.

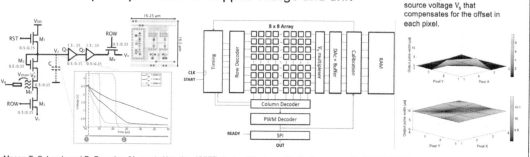

, Moser, T. S. Lande and P. Georgiou, "A novel pH-to-time ISFET pixel architecture with offset compensation," *2015 IEEE International Symposium n Circuits and Systems (ISCAS)*, Lisbon, 2015, pp. 481-484.

25   TITANICKS – A 4K ISFET array for infectious disease

- AMS 0.35um Technology
- 4368 Sensors
- pH-Time pixel
- Auto-calibration
- Trapped charge compensation
- Scalable architecture

N. Moser, J. Rodriguez-Manzano, T. S. Lande and P. Georgiou, " A Scalable ISFET
Sensing and Memory Array with
Sensor Auto-Calibration for On-Chip Real-Time
DNA Detection" *IEEE Transactions on Biomedical Circuits and Systems., 2017*

4mm

26   Vision - Next generation rapid diagnostics for infectious disease

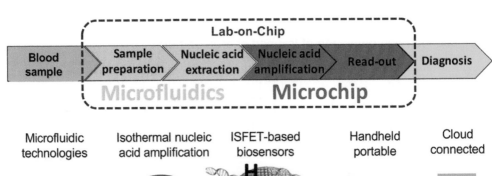

| Microfluidic technologies | Isothermal nucleic acid amplification | ISFET-based biosensors | Handheld portable | Cloud connected |

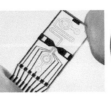

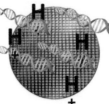

27 POC nucleic acid sensing platform

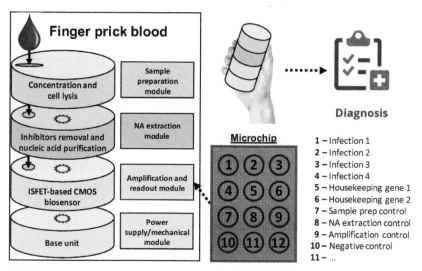

Finger prick blood

Concentration and cell lysis	Sample preparation module
Inhibitors removal and nucleic acid purification	NA extraction module
ISFET-based CMOS biosensor	Amplification and readout module
Base unit	Power supply/mechanical module

Diagnosis

Microchip

1 2 3
4 5 6
7 8 9
10 11 12

1 – Infection 1
2 – Infection 2
3 – Infection 3
4 – Infection 4
5 – Housekeeping gene 1
6 – Housekeeping gene 2
7 – Sample prep control
8 – NA extraction control
9 – Amplification control
10 – Negative control
11 – ...

4,000 ISFET array sensors; multiplex assay; <15min from sample to result

28 Isothermal DNA amplification

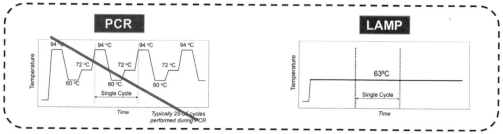

PCR

94 °C 94 °C 94 °C 94 °C
72 °C 72 °C 72 °C
60 °C 60 °C 60 °C
Single Cycle

Time — Typically 25-35 cycles performed during PCR

LAMP

63°C
Single Cycle
Time

Avoiding thermal cycling
for nucleic acid
amplification!

Time-to-positive < 10min

29 Microchips for DNA diagnostics

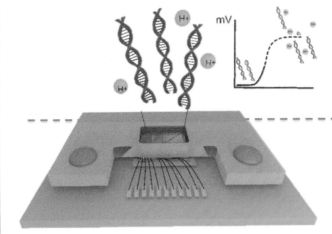

- DNA amplification
- pH variation

Molecular biology

Electronics

- Sensor for pH detection
- Microchip for ion imaging
- Platform with connectivity and portability

30 Lacewing - 4,000 pH sensors connected to your phone

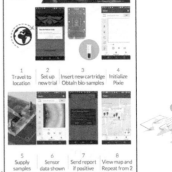

1 Travel to location
2 Set up new trial
3 Insert new cartridge Obtain bio-samples
4 Initialize Pixie

5 Supply samples
6 Sensor data shown
7 Send report if positive
8 View map and Repeat from 2

Au, A., Moser, N., Rodriguez-Manzano, J., & Georgiou, P. (2018, May). Live demonstration: a mobile diagnostic system for rapid detection and tracking of infectious diseases. In Circuits and Systems (ISCAS), 2018 IEEE International Symposium on (pp. 1-1). IEEE.

31 Detection of pathogens in under 7 minutes!

on-tube amplification

Enterobacteriaceae Resistant Gene for AMR

on-chip amplification

Miscourides, N., Yu, L.S., Rodriguez-Manzano, J. and Georgiou, P., 2018. A 12.8 k Current-Mode Velocity-Saturation ISFET Array for On-Chip Real-Time DNA Detection. *IEEE transactions on biomedical circuits and systems*, (99), pp.1-13.

Synergy between Engineering and Molecular biology to solve the challenges of LMIC countries!

32 Supersensitive detection of artemisinin resistance in malaria

P. Falciparum artemisinin-resistance

Detection <15 mins
On-chip Quantification!

artida-Cardenas, K., Miscourides, N., Rodriguez-Manzano, J., Yu, L.S., Moser, N., Baum, J. and Georgiou, P., 2019. Quantitative and rapid Plasmodium falciparum malaria diagnosis and nisinin-resistance detection using a CMOS Lab-on-Chip platform. *Biosensors and Bioelectronics*, 145, p.111678.

33 Proven on a number of pathogens to date

- **Carbapenem-resistant Enterobacteriaceae** (NDM, VIM, KPC and OXA-48)
- **Dengue typing** (DENV1-DENV4)
- **Malaria** (specific detection of *P. falciparum and* Artimisinin resistance)
- **Aspergillus** (Tandem-Repeats associated with Azole resistance: TR34 & TR46)
- **Viral vs Bacterial infections** using a 2-transcript host RNA signature

1. Rodriguez-Manzano J*, Moniri A*, Malpartida-Cardenas K, Dronavalli J, Davies F, Holmes AH, Georgiou P. Simultaneous Single-Channel Multiplexing and Quantification of Carbapenem-Resistant Genes using Multidimensional Standard Curves. **Anal Chem. 2019; in press.**
2. Yu LS*, Rodriguez-Manzano J*, Malpartida-Cardenas K, Sewell T, Bader O, Armstrong-James D, Fisher MC, Georgiou P. Rapid and sensitive detection of azole-resistant Aspergillus fumigatus by tandem-repeat loop-mediated isothermal amplification. **J Mol Diagn. 2018; in press.**
3. Malpartida-Cardenas K*, Rodriguez-Manzano J*, Yu LS, Delves MJ, Nguon C, Chotivanich K, Baum J, Georgiou P. Allele-Specific Isothermal Amplification Method Using Unmodified Self-Stabilizing Competitive Primers. **Anal Chem. 2018; 90(20):11972-11980.**
4. Miscourides N, Yu LS, Rodriguez-Manzano J, Georgiou P. A 12.8 k Current-Mode Velocity-Saturation ISFET Array for On-Chip Real-Time DNA Detection. **IEEE Trans Biomed Circuits Syst. 2018; 12(5):1202-1214.**
5. Moniri A, Rodriguez-Manzano J, Georgiou P. A framework for analysis of real-time nucleic acid amplification data using novel multidimensional standard curves. **bioRxiv. 2018 Jan 1:379180.**
6. Moser N, Rodriguez-Manzano J, Lande TS, Georgiou P. A Scalable ISFET Sensing and Memory Array With Sensor Auto-Calibration for On-Chip Real-Time DNA Detection. **IEEE Trans Biomed Circuits Syst. 2018; 12(2):390-401.**

34 A global collaboration

- AMR-CPE
- Dengue
- Tuberculosis
- Malaria
- Aspergillus
- Porcine pathogens

35 Rapid diagnostics now at the palm of your hand!

Dengue

Malaria

MRSA

36 You never fail until...

You never fail until you stop trying.

Albert Einstein

37 Acknowledgments

Collaborators:

AMR & Dengue
Alison Holmes
Sophie Yacoub

TB & Bacterial vs Viral
Mike Levin
Myrsini Kaforou
Jethro Herberg

Malaria
Aubrey Cunnington
Jake Baum

Aspergillus
Matthew Fisher
Darius Armstrong

Porcine Pathogens
Paul Langford

Dr. Pantelis Georgiou
Department of Electrical and
Electronic Engineering
Centre for Bio-Inspired Technology
http://www.imperial.ac.uk/people/pantelis

pantelis@imperial.ac.uk

@pgeorgiou_ic

CHAPTER 02

Non-invasive Bioelectronics for Personal and Precision Health Care

Alex Casson

University of Manchester

I magine having a long term condition such that you must take regular medication. Taking it is a substantial burden on you and the people who help care for you, particularly if you are elderly or suffering from cognitive impairment. What if the treatment could be released automatically, not on a predefined routine but by detecting when you actually need it? This is a powerful vision, but one which highlights the limitations of current technologies. Over the last decade exciting non-invasive sensing platforms such as the fitbit and Apple watch have enabled the capture of real-time health and lifestyle data. However, all current 'wearables' have two major limitations:

- The manufacturing is based upon highly miniaturised, but essentially conventional, microelectronics mass produced in a Silicon foundry. As such there is limited scope for the personalisation of devices from a manufacturing point of view. It also means that current devices are held in place using straps, which do not maintain a good connection to the body over long periods and which introduce significant numbers of strap and motion artefacts into the collected signals.
- Their focus has been on "Big Data" collection. Wearable sensors are now seen as "one-way streets" only for data generation. There is substantial potential for sensor systems which analyse the collected data 'on-electrode' and allow data driven personalisation of outputs. For example, monitoring physiological parameters and performing highly targeted treatment release within a few milliseconds of a trigger event, where there is not time to send the signals to a data centre for conventional analysis.

This talk will overview the confluence of several emerging technology areas aiming to tackle these limitations. By combining flexible electronics for more discrete and more personalised wearables, low power on-wearable signal processing for improving battery-life real-time data analysis for feedback and actuation, and integration with care plans; for creating next generation wearable devices, we are creating wearable devices very different to those available today.

1 Objectives

The bio- / -electrical landscape

My lab

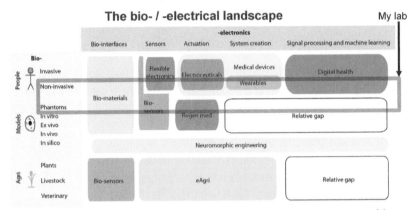

- ✓ To overview our **flexible electronics** for going beyond current wearables
- ✓ To place these in context of the **big picture** and trends in personal and precision health care
- ✓ To give examples of our **data analysis** and **closed loop** systems

2 Acknowledgements

Collaborators

University of **Kent**

cpi

Northwestern University

CHIBA UNIVERSITY

UNIVERSITY OF **TORONTO**

UNIVERSITY OF **LIVERPOOL**

Salford Royal **NHS**
NHS Foundation Trust

CARDIFF UNIVERSITY
PRIFYSGOL CAERDYᴅⅾ

sphere

An EPSRC Interdisciplinary Research Collaboration (IRC)

Team

Funders

EPSRC
Engineering and Physical Sciences Research Council

MRC Medical Research Council

BBSRC
bioscience for the future

3 Wearable monitoring

rrent wearable monitors are widely available commercially, but they also have a wide number of mitations.

4 Limitations of wearables

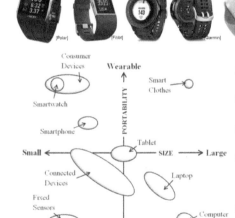

× Questionable accuracy

× Not integrated with care systems

× Limited battery life

× Modalities come from "technology push" not "clinical pull"

× Not 'invisible'

× One size fits all

[Amor, 2015]

5 eSkin Epidermal devices

| 1980s: Desktops | 1990s: Laptops | 2000s: Smartphones | 2010s: Wearables | 2020 onwards: Conformals |

✓ Tattoo-like connection to the skin: very long term

✓ Follows micro-contours of the skin: high contact area increases Signal-to-Noise Ratio

✓ Very discrete profile for social acceptability

We have a series of EPSRC projects looking at creating 'conformal' electronics as the next genera electronic platform for wearables. Central to our work is collaboration with the High Value Manufactu catapult to enable mass manufacturing.

6 Our aims

✓ Printed graphene based tracks and bio-degradable substrates for ease of disposal at end of life

✓ Screen printing for scale up. Inkjet printing for personalisation

✓ Roll-to-roll scale up manufacturing

✓ 3D pop-up antennas

✓ <30 μW power budget for 3m range RFID wireless powering

✓ Sensing: ECG, EMG, EEG, IMU, Temp, Strain, Pressure, pH

✓ High data rates (5G transmission)

Conductive smart textiles   Temporary tattoo sensors

At present we are looking at relatively 'standard' sensing modalities in order to establish the manufactu platform. We will then add in more and more modalities over time.

7 System examples

Silver ink screen printed on PET

Significant complexity in board

Graphene based material ink on paper

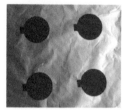

Commercial graphene based material ink screen printed onto Cu foil as battery anodes

---

8 EEG recording

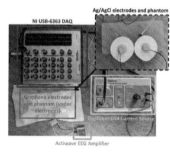

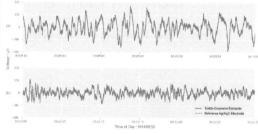

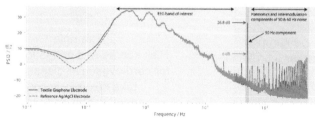

is example shows EEG "brainwaves" being recorded from a body phantom using printed graphene on a otton substrate. It's just one illustration of flexible electronics for next generation wearables.

The bigger picture

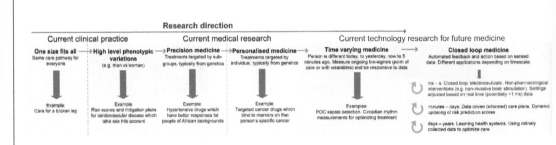

Zooming out, let's take a look at the "Big Picture" and where we're heading with this.

The bigger bigger picture (I)

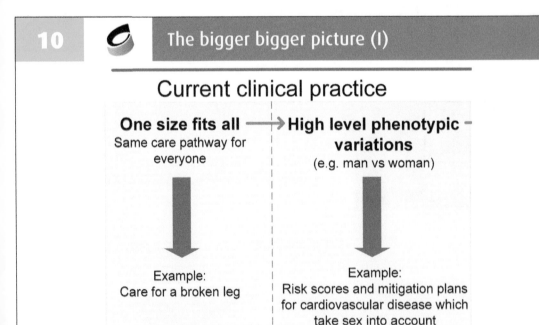

Current clinical practice is one size fits all.

11 The bigger bigger picture (II)

Research direction

Current medical research

 Precision medicine **Personalised medicine**

Treatments targeted by sub-groups, typically from genetics

Treatments targeted by individual, typically from genetics

Example:
Hypertensive drugs which have better responses for people of African backgrounds

Example:
Targeted cancer drugs which bind to markers on that person's specific cancer

ere's huge amounts of medical research going into "precision" and "personalised" medicine. sically these are genetics based. We can target interventions based upon the genetic sub-groups people belong to, knowing that there are treatments which work in some groups but not in others.

12 The bigger bigger picture (III)

Current technology research for future medicine

Time varying medicine ⟶ **Closed loop medicine**

Person is different today, to yesterday; now to 5 minutes ago. Measure ongoing bio-signals (point of care or with wearables) and be responsive to data

Automated feedback and action based on sensed data. Different applications depending on timescale.

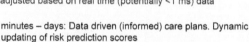

ms – s: Closed loop 'electroceuticals'. Non-pharmacological interventions (e.g. non-invasive brain stimulation). Settings adjusted based on real time (potentially <1 ms) data

minutes – days: Data driven (informed) care plans. Dynamic updating of risk prediction scores

Examples:
POC sepsis detection. Circadian rhythm measurements for optimizing treatment

days – years: Learning health systems. Using rotinely collected data to optimize care

ir genetics are nearly completely fixed though. The next step is to use continuously generated – from point of care (POC) devices in clinical ngs and wearables outside of clinical settings to ct that fact people are different day to day. This then allows closed loop interventions using all this data to change how people interact with the care system. Depending on the timeframe for closing the loop the area is given different names.

13 Healthcare provision today

No use of wearable / longitudinal data to inform appointments, treatment

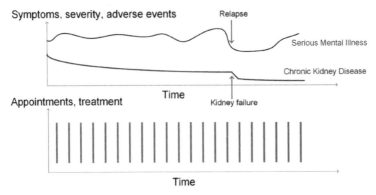

× Appointments/treatments are regular, fixed. No dynamic care planning

× No use of additional information: wearables, smartphones, digital footprint

× Wearable data on proprietary cloud isn't sufficient for health and social care impact

For example, this is one targeted end point of wearable devices: allowing data informed care plans.

14 Where we want to get to

Enable more out-of-the-clinic data driven care

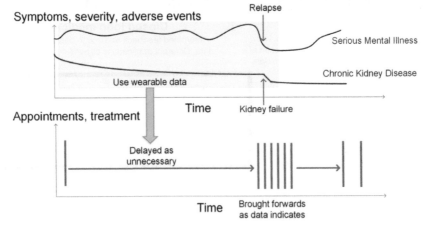

✓ Significant savings in health care resources, unnecessary appointment

✓ More timely and so more effective appointments, treatment and similar

✓ Reduced burden on person

15 **Requirements**

Much more than a new wearable device or new sensing method

- Healthcare economics to ensure is cost effective.
- Risk analysis / assurance case to ensure is safe.
- Patient / public involvement to ensure is usable.
- Data integration to allow different parts to interact.

- Detailed electronically executable care plans.
- Data responsive care plans.
- Risk prediction from electronic health record to guide wearable sensing.

- Dynamic risk prediction from wearable / longitudinal data.
- Integration of different models for multi-dimensional risk prediction.
- Risk prediction score to guide wearable sensing and care plan.

Not necessarily new wearable sensors, new sensing modalities in existing
sensors, or improving sensitivity/selectivity/robustness/accuracy

16 **UK care pathway**

Formulism of care pathway for home blood pressure monitoring

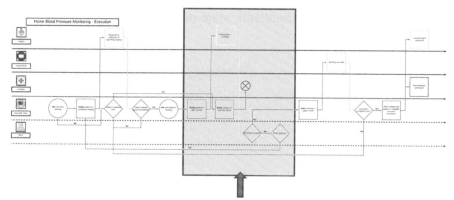

Automated Blood Pressure (BP) measurements
Only take recording once user is sat for 5 minutes, as
required in NICE guidelines

ur EPSRC "Wearable clinic" project is tackling some of these questions, investigating blueprints for creating
data executable care plans.

17 On-wearable processing

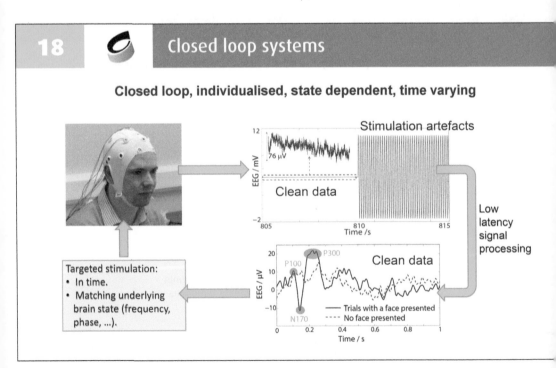

Wearable sensor node

Biological world — Electronic world

Brainwaves · Heart rate · ... · Gait

Interface / Electrode

Signal conditioning

Online signal processing

Wireless link

Power delivery (Battery or energy harvesting)

Feedback / treatment

Smartphones PCs Internet ... Offline signal processing

mHealth Big Data Internet of Things

✓ Reduce system power.
✓ Increase functionality.
✓ Better quality recordings.
✓ Minimise system latency.

✓ Reduce amount of data to analyse.
✓ Reliable operation over unreliable wireless.
✓ Enable closed loop: recording – stimulation.
✓ Data redaction for privacy.

My lab also has a wide range of work on real-time on-wearable processing of data. Historically the aim of this was power reduction – reducing the amount of data that had to be transmitted from the wearable. Today the aims are to enable clos loop systems, particularly record-stimulate or and in keeping data local to the person for priva preservation.

18 Closed loop systems

Closed loop, individualised, state dependent, time varying

Stimulation artefacts

76 µV

Clean data

EEG / mV

805   810   815
Time /s

Low latency signal processing

Clean data

P100   P300

N170

EEG / µV

— Trials with a face presented
---- No face presented

0   0.2   0.4   0.6   0.8   1
Time / s

Targeted stimulation:
• In time.
• Matching underlying brain state (frequency, phase, ...).

We have a particular focus on non-invasive brain stimulation, using electricity, light, sounds, and ot modalities. At the heart of this is low latency signal processing.

19 **Platforms**

Non-invasive closed loop stimulation

Sound for sleep engineering	Electrical stimulation	Visual stimulation
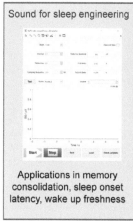	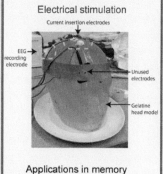	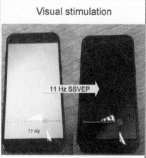
Applications in memory consolidation, sleep onset latency, wake up freshness	Applications in memory consolidation, post-stroke, Parkinson's	Applications in chronic pain

Many additional opportunities. Particularly for invasive systems

We have a number of different systems being used in applications such as sleep engineering and in chronic pain. The aim is to move beyond devices which "only collect data".

---

20 **Summary**

Personalised medicine: in manufacture and in data driven responses

Flexible electronics

Data driven care plans

Closed loop stimulation

Flexible+care plan+closed loop

@a_casson
alex.casson@manchester.ac.uk
www.eee.manchester.ac.uk

We have a number of projects across quite a broad spectrum. Any one of these parts in isolation is important, but combined they really do give wearable devices which are a step change from what we have today.

Beyond "Wearable" Micro/Nano-Scale Bio-Sensors for Selected Applications in Healthcare

Ian Underwood

School of Engineering
University of Edinburgh

Contents

This seminar showcases the results of two research projects both involving micro/nano-scale bio-sensors and both located on the periphery of wearable technologies. The first part of the seminar reports on some aspects of a large multidisciplinary project funded by an EPSRC Programme Grant entitled Implantable Microsystems for Personalised Anti-Cancer Therapy, EP/K034510/1 (https://www.impact.eng.ed.ac.uk/home). Miniature sensors have been developed for implantation to measure oxygen and pH in solid tumours in order to assist with optimisation of the delivery of radiotherapy. The sensors have been shown to function successfully in sheep with lung cancer. The latter part of the seminar describes substantial progress towards a sensing continuous capability integrated in a smart electrochemical contact lens that will allow non-invasive monitoring of biomarkers that may otherwise require to be measured in blood. A real-time "video" shows high concentration fluid flowing through a model eye with the ESCL measuring varying concentration at each electrodes over time. Input, flow and drainage of an incoming higher concentration fluid can be seen. This research constitutes part of a PhD project within the EPSRC Centre for Doctoral Training in Intelligent Sensing and Measurement, EP/L016753/1.

1 — Two Applications; two projects

- **Precision Radiotherapy**
- Implantable (Wearable?) Sensors of O_2 and pH
- *"Implantable Microsystems for Personalised Anti-Cancer Therapy (IMPACT)"*

- **Continuous Non-Invasive Monitoring of Selected Biomarkers**
- Sensors Integrated in Contact Lens to make Measurements of Biomarkers in Tear Fluid
- *"Electrochemical Smart Contact Lens (ECSL)"*

2 — Contents

IMPACT
- Introduction to IMPACT
- Miniature Clark Oxygen sensors
- Miniature ISFET Senors for pH measurement

Electrochemical Smart Contact Lens
- Introduction to smart contact lenses
- Design of the ESCL
- Demonstration of the sensing technique

IMPLANTABLE MICROSYSTEMS FOR PERSONALISED ANTI-CANCER THERAPY

3 **Cancer - Statistics**

- 360,000 new cancer diagnoses in the UK every year
  - Almost 1000 every day (2013-2015)
  - One new diagnosis every 2 minutes
  - Annual cost – including loss of productivity … £18.3 billion

- Radiotherapy (RT) is a key treatment
  for many solid cancers

- IMPACT aims to
  - Target RT for maximal effectiveness
  - Measure success of RT in terms of cancer cell kill

---

4 **A highly multi-disciplinary collaboration**

Electronic Engineers
- Device designers and technologists, IC designers, Microfabrication technologists

Chemists
- Electrochemists, molecular synthetic chemists

Medics
- Clinicians, cancer specialists

Veterinarians
- Cancer specialists

Social scientists
- Policy advocates, ethicists

More than 35 researchers in total !!!

5 IMPACT cast list

ENG. Prof Alan Murray (Principal Investigator), Dr Ewen Blair, Dr Camelia Dunare, Dr Brian Flynn, Dr Chandrasekaran Gunasekaran, Dr Jamie Marland, Dr James Meehan (HWU), Dr Martin Reekie, Dr Stewart Smith, Dr Jon Terry, Dr Andreas Tsiamis, Prof Ian Underwood, Prof Anthony Walton, Prof Steve McLaughlin (HWU), Ricky Guerrero, Liyu Huang, Nadira Jamil, Ahmet Ucar.

CHEM. Prof Mark Bradley, Prof Andy Mount, Dr Nicos Avlonitis, Dr Eva Gonzalez Fernandez, Dr Matteo Staderini, Dan Norman

CLIN. Prof David Argyle, Dr Carol Ward, Prof Ian Kunkler, Dr Duncan McLaren, Prof Edwin van Beek, Dr Bill Nailon, Mark Gray

VAL. Prof Joyce Tait, Dr Andrew Watkins, Dr Theresa Ikegwuonu, Dr Ann Bruce, Dr Gill Haddow

6 What is "Hypoxia"?

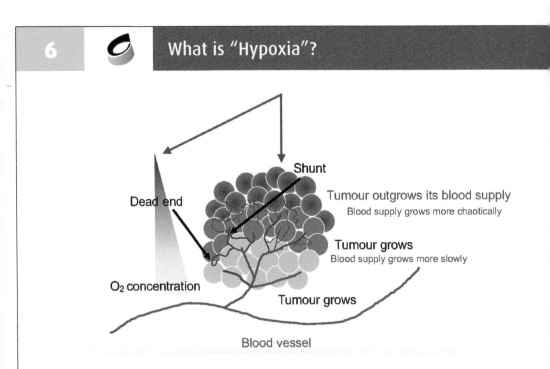

Shunt

Dead end

Tumour outgrows its blood supply
Blood supply grows more chaotically

Tumour grows
Blood supply grows more slowly

O_2 concentration

Tumour grows

Blood vessel

7 Radiotherapy

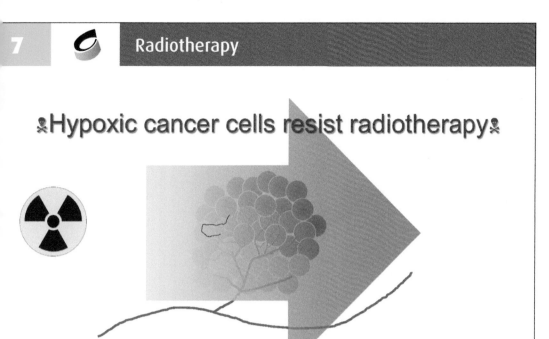

☠Hypoxic cancer cells resist radiotherapy☠

8 Personalised anti-cancer therapy?

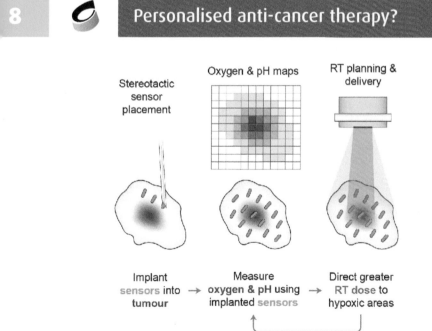

Stereotactic sensor placement

Oxygen & pH maps

RT planning & delivery

Implant sensors into tumour → Measure oxygen & pH using implanted sensors → Direct greater RT dose to hypoxic areas

9 — Miniature implantable O_2 sensor

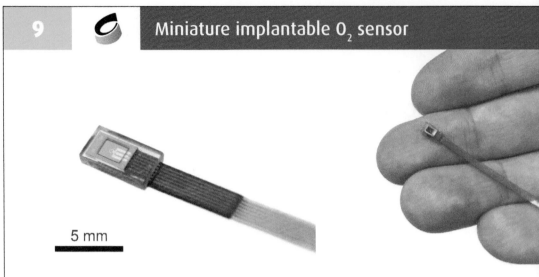

5 mm

- **Tissue O_2 sensor** for biomedical applications
- Fabricated on silicon using **CMOS-compatible processes**
- Encapsulated in **biocompatible packaging**

10 — Sensing oxygen using electrochemistry

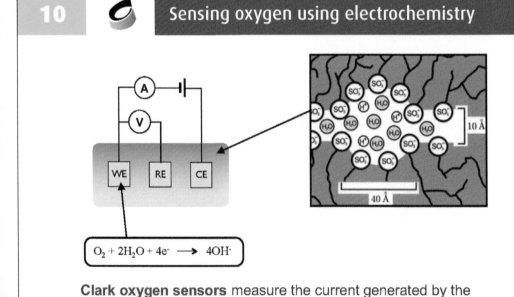

$$O_2 + 2H_2O + 4e^- \longrightarrow 4OH^-$$

Clark oxygen sensors measure the current generated by the **electrochemical reduction** of **oxygen** at a platinum cathode

11 ## O$_2$ sensor - Bench testing

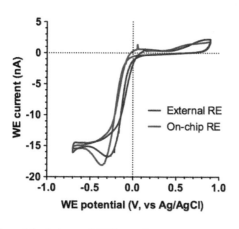

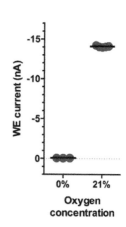

On-chip internal RE performance comparable to commercial external RE

Responds to changes in O$_2$ concentration in solution

12 ## Sheep lung cancer model

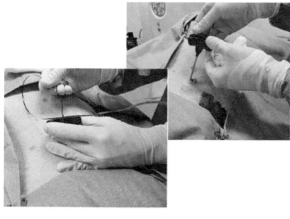

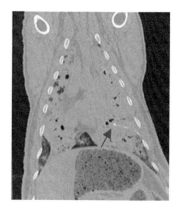

- Preclinical testing of **sensor performance** in sheep with lung cancer
- **Highly translational:** similar biology and surgical techniques to human
- Results encouraging!

13   Monitoring after bowel surgery

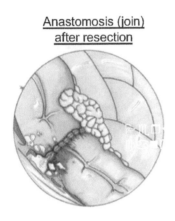

Anastomosis (join)
after resection

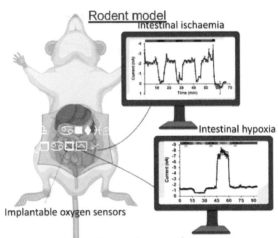

Rodent model

Intestinal ischaemia

Intestinal hypoxia

Implantable oxygen sensors

- Alternative application: detection of **anastomotic leakage** after surgery
- Could provide **"early warning"** of leak developing, allowing timely intervention before irreversible deterioration

---

14   Implantable ISFET pH sensors

Microfabricated pH sensors can be produced using ion-sensitive field-effect transistor (ISFET) technology

ISFETs are analogous in construction and principle of operation to a conventional MOSFET.

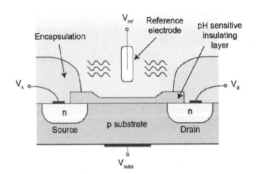

Schematic cross-section showing the structure of a pH-sensitive ISFET

15 **CMOS ISFET Sensors: On-chip post-process**

Why?

- Less costly to prototype

- Non standard foundry processes

- Add extra functionality

- Increase sensitivity

- Reduce signal attenuation

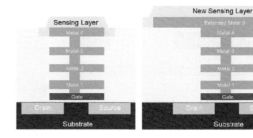

Foundry Design   **Modified Design**

16 **Post-processing: Designs and tool compatibility**

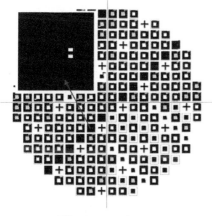

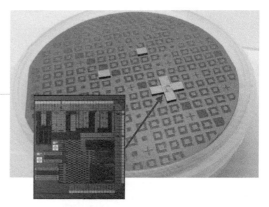

Photomask Design

Product Carrier Wafer

17 — Post processing: Foundry de-passivation

1. Foundry chip
2. Photolithographic pattern
3. Passivation RIE

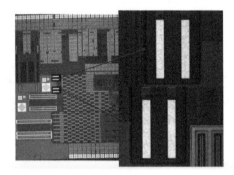

18 — Post processing: Metallisation

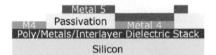

1. Photolithographic pattern
2. Metallisation
3. Lift-off process

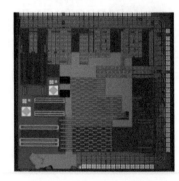

19 Post-processing and more

New sensing layer deposition

- PECVD, sputtering, metal anodization

- Packaging, wire bonding and encapsulation

Other post-processing designs

- On-chip reference electrodes

- Complete sensor system on CMOS with instrumentation and readout

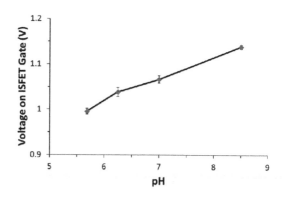

20 CMOS ISFET sensors for IMPACT: Summary

4 ASIC generations

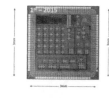

Multiple instrumentation and sensor designs for:

- Development & characterisation

- *In vivo* research

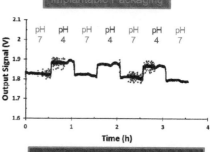

Implantable Packaging

Continuous real-time pH monitoring

21 IMPACT sensors summary

- Microscale sensors on silicon developed and characterized
- For oxygen, pH and some biomarkers
- Packaging developed and robustness demonstrated

- Initial in-vivo testing completed
- (On-chip circuits also designed)
- but publications still pending
- Hence only some preliminary results shown here

ELECTROCHEMICAL SMART CONTACT LENS (ECSL)

22 What are smart contact lenses?

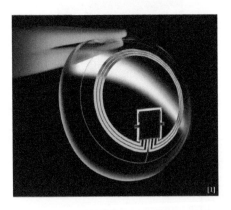

- Contact lenses with electronic components
- Uses: glucose sensing for diabetes; pressure sensing for glaucoma; electronic focusing and optical zoom

23 Electrochemical smart contact lens (ESCL)

What is an ESCL?

- Contact lens with integrated electrochemical sensors
- Measure substances (such as glucose) *non-invasively* in the tears

5 mm

Advantages

- Continuous, unobtrusive sensing is *more comfortable* – and *more informative* – than getting a blood test
- Many biomarkers may be measured at the same time, such as alcohol, hormones, glucose and others

24 Three important factors in design

Material Design

- Comfortable and safe in the eye
- Flexible
- Transparent
- Oxygen permeable
- Biocompatible

Sensor Design

- High performance
- Select for specific markers
- Flexible and robust during bending

Algorithmic Design

- Low power consumption
- Minimal circuitry and signal processing
- Micro-sized sensor system – thinner than a human hair!

25 Our ESCL - Material design

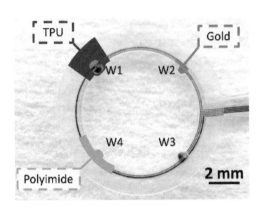

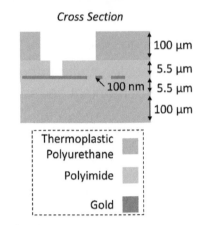

Cross Section

- Flexible, biocompatible polymer design
- Embedded in hydrogel after fabrication

26 Our ESCL - Sensor design

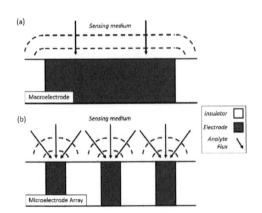

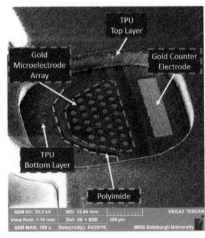

27 Our ESCL - Algorithmic design

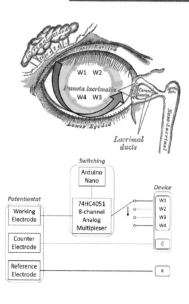

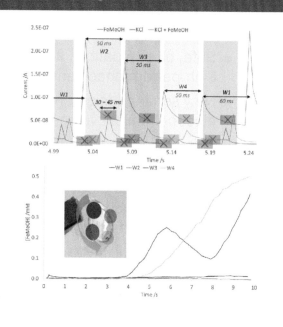

28 Results

Video: Concentration measured during flow in a model eye

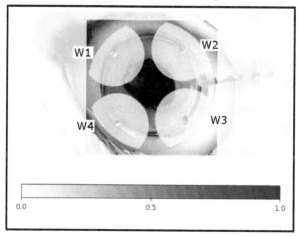

29 ESCLs - Summary

- Safe, biocompatible, flexible ESCLs can be made with multiple high-performance sensors integrated
- New sensor design and measurement techniques produce electrochemical videos of substances flowing across the eye
- This design and methodology helps improve our understanding of chemical measurements made in SCLs
- With this technology, will be able to measure various "health parameters" continuously non-invasively and unobtrusively, in a way that has not previously been possible
- Next steps will involve design and inclusion of integrated circuitry

30 Acknowledgements

The authors gratefully acknowledge
- **UKRI / EPSRC EP/K034510/1**
  - Implantable Microsystems for Personalised Anti-Cancer Therapy, £4.4M, 5/2013 – 5/2019
- **UKRI / EPSRC EP/L016753/1**
  - Matthew Donora's studentship was funded through
  - EPSRC Centre for Doctoral Training in Intelligent Sensing and Measurement, £4.8M, 5/2014 – 10/2022.

31 Further information

The IMPACT web site contains all publications arising from the IMPACT project
https://www.impact.eng.ed.ac.uk/home

ECSL publications to date include –

- **Electronic Contact Lens for Senses beyond Sight,** Matthew Donora, Ian Underwood, SID Symposium Digest of Technical Papers, Volume 50, Issue 1, https://doi.org/10.1002/sdtp.13094
- **Spatiotemporal Electrochemistry on Flexible Microelectrode Arrays: Progress Towards Smart Contact Lens Integration.** Donora, Matthew; González-Fernández, Eva; Quintero, Andrés Vásquez ; de Smet, Herbert; Underwood, Ian. In: Sensors and Actuators B: Chemical, Vol. 296, 126671, 10.2019. https://doi.org/10.1016/j.snb.2019.126671
- **Spatiotemporal Electrochemical Sensing in a Smart Contact Lens,** Donora, M., Quintero, A. V., de Smet, H. & Underwood, I., E-pub ahead of print - 6 Oct 2019, In: Sensors and Actuators B: Chemical. https://doi.org/10.1016/j.snb.2019.127203

Novel Circuits and Systems Technologies to Monitor the ECG in Fetuses and Neonates Non-Invasively

Elizabeth Rendon-Morales

University of Sussex

In this talk, I will present our work on the development of circuits and systems technologies based on electric potential sensors to non-invasively detect cardiac electrical activity of babies during and after pregnancy.

We will describe our design to monitor both: the maternal and foetal heart activity during pregnancy. We will show the suitability of our technology to monitor the foetal electrocardiogram (ECG) starting at week twenty in a study conducted for ten weeks. Importantly, ECG data is presented without any post processing given that our technology eliminates the requirement of signal conditioning algorithms such as having to un-mix both, the maternal and foetal ECG.

Then, we will present our novel smart mattress design to measure the babies' ECG and respiration rate immediately after birth. We will present proof of concept tests to demonstrate the potential of our technology to provide a quick and reliable application to obtain ECG readings in less than 30 seconds.

The benefits of our technology are: during pregnancy, to potentially determine heart related congenital disorders. After pregnancy, to assist the neonatal staff with those neonates requiring resuscitation though the display of real time ECG information required to assess the success such procedure.

1 Presentation outline

- ### Our research (brief history)
- ### The technology
- ### ECG monitoring:
  - during pregnancy
  - after pregnancy
- ### Conclusions
- ### Future outlook

2 Our research

Danio Rerio (Zebra fish)

3 The challenge

- Electric field detection of cardiac and the neuronal activity in zebrafish for drug discovery applications

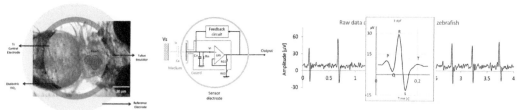

For 3/5 DPF The mean peak amplitude is
22.5 ± 3 µV/ 28.3 ± 3 µV.

E. Rendon-Morales, R.J. Prance, H. Prance, R. Aviles-Espinosa, "A Novel Non-invasive
Biosensor Based on Electric Field Detection for Cardio-electrophysiology in Zebrafish
Embryos," Procedia Technology, 27, pp, 242-243 (2017).

4 The technology

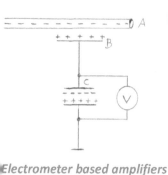

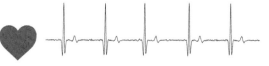

- A) Negatively charged Rod
- B) Metal plate
- C) Coulombmeter

Electrometer based amplifiers

5 Electrometer based amplifiers

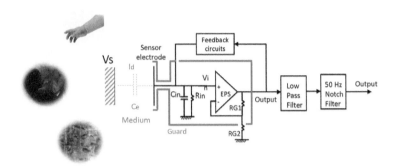

- Sample coupling: High input impedance up to 10^{15} Ω
- Avoid altering measurement: low Input capacitance ~ 1pF
- Low noise reference output < 30 nV/Hz $^{1/2}$

ECG MONITORING... DURING PREGNANCY...

6 Foetal monitoring

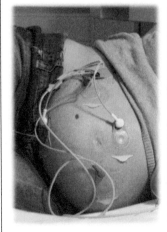

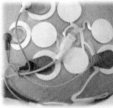

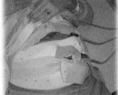

- An Effective method to ensure the wellbeing a baby.

- Allow clinicians to make informed decisions.

- Required for gaining understanding on medical complications.

- Required for women undergoir high risk pregnancies foetal electrocardiogram monitoring.

- Current solutions…

7

Current technologies for fECG detection (I)

	Philips Avalon CL	Pregnabit	Meridian M110	Monica AN24	Monica Noviipatch
Clinic Based	✓		✓	✓	✓
Home based		✓		?	
Complex	✓	✓	✓		✓
Pressure required	✓	✓			
Requires gel/ skin preparation	✓	✓			✓
Ultrasound (CTG)	✓	✓			
fECG			✓	✓	✓
Complex algorithms for HR separation	✓	✓	✓	✓	✓

8

Current technologies for fECG detection (II)

	Philips Avalon CL	Pregnabit	Meridian M110	Monica AN24	Monica Noviipatch
Clinic Based	✓		✓	✓	✓
Home based		✓		?	
Complex	✓	✓	✓	✓	✓
Pressure required	✓	✓			
Requires gel/ skin preparation	✓	✓			✓
Ultrasound (CTG)	✓	✓			
fECG			✓	✓	✓
Complex algorithms for HR separation	✓	✓	✓	✓	✓

9   Global fECG requirements

- FECG tests are compulsory/strongly recommended in many countries, even for low-risk pregnancy

- Home-based FECG is an emerging technology.

- The prominent companies (Philips, GE Healthcare and Siemens Healthcare) are much less dominant in this area.

- Monica Health in the UK just acquired by GE.

Global market share in foetal monitoring MartketsandMarkets (2015) 'Fetal Monitoring Market Report to 201[...]

10   Mother (I)

What is the M◯THER project about?

- High-risk pregnancies require regular fECG monitoring to ensure the baby wellbeing.

- Most FECG solutions only available onsite at hospitals and are complex.

- Combining printed electronics and EPS embodied in a garment, our technology aims to provide a continuous and reliable home-based fECG monitoring to ensure safe foetal development.

EPS Patent no. US8264247B2, E.Rendon-Morales et. al. AIP Advances 8 (105115) (2018)

11 Mother (II)

https://uk.reuters.com/video/2019/01/28/stress-free-way-to-listen-to-your-unborn?videoId=508723297

12 Initial trial using EPS technology

Protocol

- Singleton pregnancy

- Calm atmosphere

- Mother lying down in left/right lateral position

- Baby presentation identified

- 2 electrodes for the baby (below umbilicus and above the bikini area)

- Recordings for 10 weeks

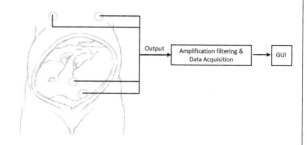

EPS Patent no. US8264247B2, E.Rendon-Morales et. al.," Non-invasive recordings of fetal electrocardiogram during pregnancy using electric potential sensors," AIP Advances 8 (105115) (2018)

13 Results (I)

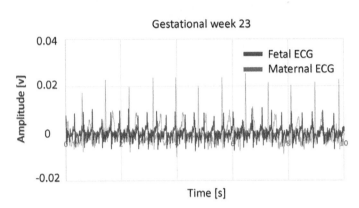

Gestational week 23

EPS Patent no. US8264247B2, E.Rendon-Morales et. al.," Non-invasive recordings of fetal electrocardiogram during pregnancy using electric potential sensors," A Advances 8 (105115) (2018)

14 Results (II)

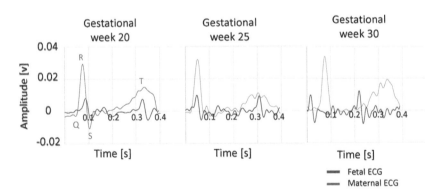

EPS Patent no. US8264247B2, E.Rendon-Morales et. al.," Non-invasive recordings of fetal electrocardiogram during pregnancy using electric potential sensors," AIP Advances 8 (105115) (2018)

15 Results (III)

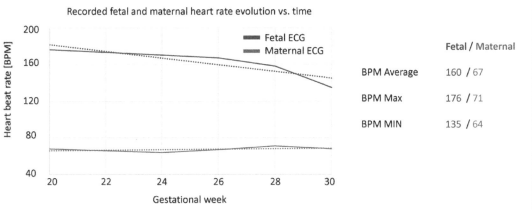

Recorded fetal and maternal heart rate evolution vs. time

Legend: Fetal ECG, Maternal ECG

	Fetal / Maternal
BPM Average	160 / 67
BPM Max	176 / 71
BPM MIN	135 / 64

EPS Patent no. US8264247B2, E.Rendon-Morales et. al.," Non-invasive recordings of fetal electrocardiogram during pregnancy using electric potential sensors," AIP Advances 8 (105115) (2018)

...FTER PREGNANCY...

16 Neosense (I)

What is the Neo-sense project about?

If a baby does not start breathing, its heart rate will drop and the circulation of blood carrying oxygen to the organs will be seriously affected.

Requires prompt actions i.e. no preparation

During birth, the attending neonatal staff manually listen to the baby's heart and count the heart rate.

Requires no interruptions

Research on printed electronics and EPS sensing embodied in a smart mattress.

17   Neosense (II)

Protocol

- Young infant at rest

- Calm atmosphere

- Parents lay down the young infant on top of the sensor integrated mattress

- 2 electrodes system (making contact with the infants back

- 10 five minutes recordings

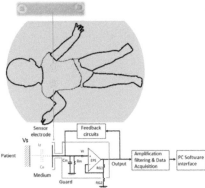

EPS Patent no. US8264247B2, Aviles-Espinosa, R, Rendon Morales, E, et al. "Neo-Sense: a non-invasive smart sensing mattress for cardiac monitoring of babies". IEEE 2019

18   Neosense (III)

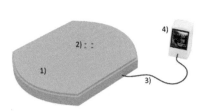

1. Modified neonatal mattress
2. Screen printed Electric Potential Sensor electrodes
3. Flexible shielded connector cab
4. Head unit – display, electronics battery unit

EPS Patent no. US8264247B2, Aviles-Espinosa, R, Rendon Morales, E, et al. "Neo-Sense: a non-invasive smart sensing mattress for cardiac monitoring of babies". IEEE 2019

O. Anton, et al, "Heart rate monitoring in newborn babies: a systematic review," Congress of joint European Neonatal Societies Maastricht September 2019.
O. Anton, et al, "Neo-Sense - a novel sensor to detect neonatal heart rate: Qualitative study on healthcare staff," JENS 2019.

19   Results (I)

ECG and Respiration vs. time

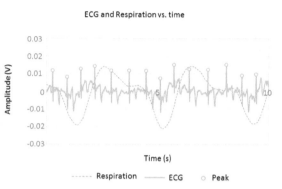

Time (s)

------ Respiration   ——— ECG   ○ Peak

HR

- Neosense: **93.12 ± 0.8 bpm**
- PO: **92.2 ± 0.7 bpm**

Respiration rate:

- Neosense: **16.3 ± 0.26 bpm**
- Reference studies: **lower limit**

EPS Patent no. US8264247B2, Aviles-Espinosa, R, Rendon Morales, E, et al. "Neo-Sense: a non-invasive smart sensing mattress for cardiac monitoring of babies". IEEE 2019

20   Results (II)

Single ECG pulse extracted from Raw data

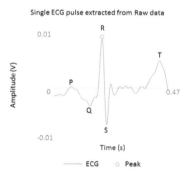

Time (s)

——— ECG   ○ Peak

EPS Patent no. US8264247B2, E.Rendon-Morales et. al.," Non-invasive recordings of fetal electrocardiogram during pregnancy using electric potential sensors," AIP Advances 8 (105115) (2018)

21 Future outlook...

- Take our prototypes development to the next stage (TRL 4-5) towards starting a clinical study.

- Employ encapsulation techniques with silver/carbon inks for sensor integration.

- Wireless transmission integration.

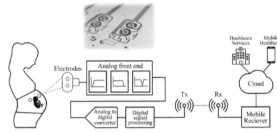

22 Conclusions

- Proof of principle demonstrations of our technology for recording: ECG during pregnancy and after pregnancy non invasively.

- EPS technology does not require skin preparation (maternal fluids)

- Recordings can be carried out using the same equipment as the pregnancy evolves

- Additional electrodes can be added to avoid having to locate the foetal presentation/ account for movement of the neonate

- Maternal and Foetal signals are not mixed

- Additional data can potentially be extracted though advanced algorithms

CHAPTER 05

Remote Sensing of Forces in Orthopaedic Implants:
An Overview

Stephen JG Taylor

University College London

The direct measurement of forces from within orthopaedic implants is seen as a 'gold standard' and is useful for furthering our knowledge of the biomechanics of the musculoskeletal system as well as for implant design and testing, physiotherapy, fracture fixation and for validating mathematical models so that they are empowered to their best advantage. This talk will present these objectives and then focus on the challenges in the design of such implantable devices, and in particular the electrical design of some implants designed and built at UCL Stanmore over the last 30 years, both the implanted and external circuits. Options for calibrating such devices will be presented, together with a selection of measurements recorded from a few instrumented implants, showing broad agreement between measured forces and those determined from even simple mathematical models during human gait. Factors affecting longevity of such implants in vivo will be discussed.

1

Why is force measurement important within the musculoskeletal system?

Musculoskeletal health is determined both by biology and by mechanics

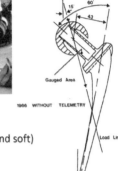

How large are joint forces in vivo?
We need to know for:
- Basic understanding of how the MS system works (hard and soft)
- Implant design and testing
- Rehabilitation
- Model validation

Fig. 2. Rydell strain gauged prosthesis (1966)

2

Simple musculoskeletal models can be developed to find internal bone forces

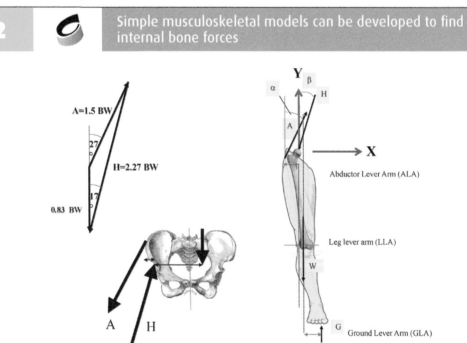

3 Types of levers in human body

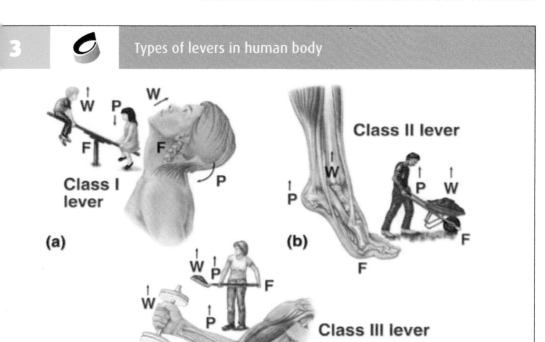

Class I lever

(a)

Class II lever

(b)

Class III lever

4 Forces acting on the femoral head-prosthesis

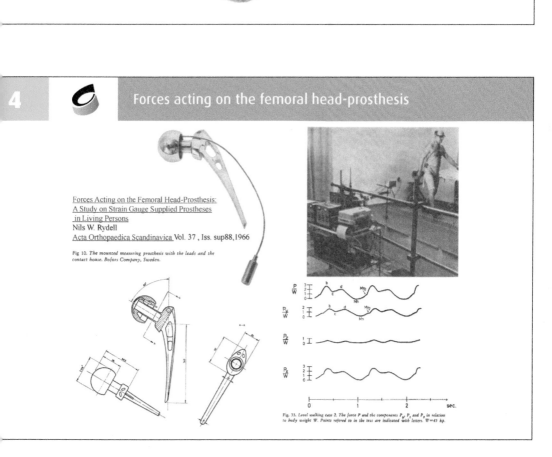

Forces Acting on the Femoral Head-Prosthesis:
A Study on Strain Gauge Supplied Prostheses
in Living Persons
Nils W. Rydell
Acta Orthopaedica Scandinavica Vol. 37 , Iss. sup88,1966

Fig 10. The mounted measuring prosthesis with the leads and the contact house. Bofors Company, Sweden.

Fig. 33. Level walking case 2. The force P and the components P_x, P_y and P_z in relation to body weight W. Points referred to in the text are indicated with letters. W=43 kp.

5 Why measure forces from within the musculoskeletal system? (I)

1 To find functional loading for implant design and testing, and for defining standards. Wear & fatigue test machines need to apply physiologically representative forces.

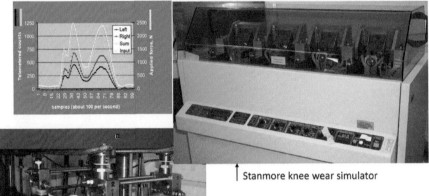

Stanmore knee wear simulator

Apply cyclic loading at a physiological rate (1-5Hz)

Hydraulic fatigue testing machine

---

6 Why measure forces from within the musculoskeletal system? (II

Offset loading on fixation

2 To see how forces are generated and distributed throughout the musculoskeletal system.

To aid our understanding of Wolff's Law, 1892 (architecture and mass of bone is suited to mechanical demand).

This enables us to understand the processes leading to the wear and loosening of joints.

replaced bone

load line (assumed)

offset

point of maximum bending stress

transection point

Correlation of radiographic and telemetric data from massive implant fixations. J Biomech **39**, 2006. Shah, Taylor, Hua.

7 Why measure forces from within the musculoskeletal system? (III)

3 To monitor fracture healing

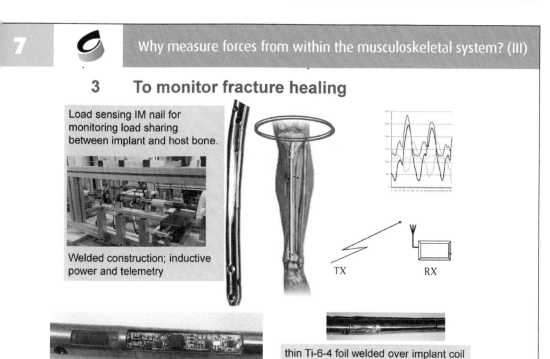

Load sensing IM nail for monitoring load sharing between implant and host bone.

Welded construction; inductive power and telemetry

TX   RX

thin Ti-6-4 foil welded over implant coil

8 Bone healing monitor for fracture repair

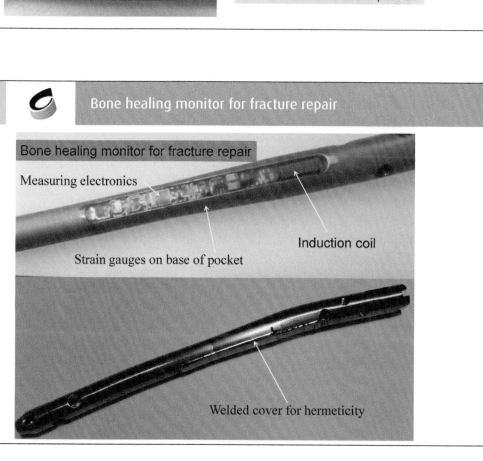

Bone healing monitor for fracture repair

Measuring electronics

Induction coil

Strain gauges on base of pocket

Welded cover for hermeticity

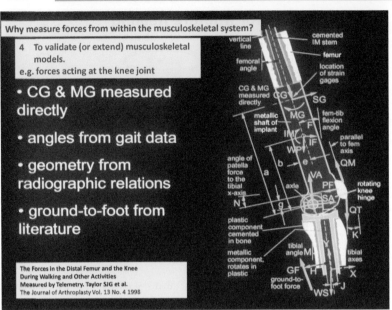

Power supply: induction v battery

Data modulation type and rate

Type of strain gauge: foil, semiconductor, thin film, thick film, piezoelectric

Location of instrumentation: inside vs. outside implant

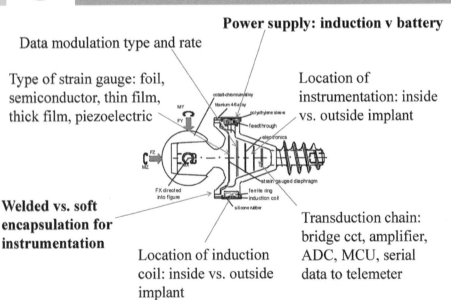

Welded vs. soft encapsulation for instrumentation

Location of induction coil: inside vs. outside implant

Transduction chain: bridge cct, amplifier, ADC, MCU, serial data to telemeter

11 Joint force telemetric links in man: parameters by various groups

Parameters	Research groups (1966 – 2010)						
Implants used *in vivo*: Hip, knee, spine, shoulder, femur.	Rydell (Hip)	English (Hip)	Davy (Hip)	Mann (Hip)	Bergmann (all)	D'Lima (Knee)	Taylor (Femur, shoulder)
Implant Power source	wire	impl battery		inductive coupling			
Method of Telemetry	wire	radio tx from implanted antenna					ind coup
Type of strain gauge	foil	foil	semiconductor			foil	thin film
Sealing of gauges	silicone rubber			welding			
Site of telemeter	-	tissues		inside prosthesis			
Coil ins/outside metal	-	-	-	outside	inside	inside	outside
Number of subjects	2	1	3	2	>50 *	2 *	5 *
Implant-years *in-vivo* data to date	0.5 **	0.1	0.25	8	200?	5?	9

*Reported thus far; further implantations likely.

**Includes rehabilitation period. Measurements terminated at 6 months post-op.

12 Options for an inductive power supply and telemeter (I)

- Implant coil welded inside metal implant

- Requires telemetry by radio antenna in ceramic part (Bergmann) or plastic part (D'Lima), both using an electrical feedthrough

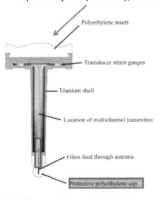

Polyethylene insert
Transducer strain gauges
Titanium shell
Location of multichannel transmitter
Glass feed through antenna
Protective polyethylene cap

Knee joint forces: prediction, measurement, and significance. D'Lima et al. Proc Inst Mech Eng H. 2012 Feb; 226(2): 95–102.

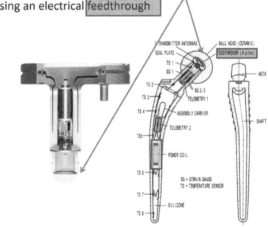

https://jwi.charite.de/en/research/reserach_groups/loading_and_movement/instrumented_implants

13 Options for an inductive power supply and telemeter (II)

- Implant coil in soft encapsulant outside implant
  - Integral with implant (e.g. Taylor)

  - Remote from implant; umbilical cable used (e.g. Burny)

- Telemetry by radio antenna (e.g. Burny);

  or Impedance modulation using same pair of coils (e.g. Taylor)

Passive signalling by impedance modulation

implant   detector

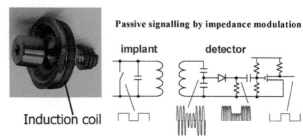

Concept, design and fabrication of smart
orthopedic implants. Burny et a., Medical
Engineering & Physics 22 (2000) 469–479

Induction coil

SCIMITAR: Subject-carried implant monitoring inductive telemetric ambulatory reader for remote
data acquisition from implanted orthopaedic prostheses. Hao, Gorjon, Taylor. Medical Engineering &
Physics, Volume 36, Issue 3, March 2014, Pages 405-411

---

14 SCIMITAR: Subject-Carried Implant Monitoring Inductive Telemetric Ambulatory Recorder

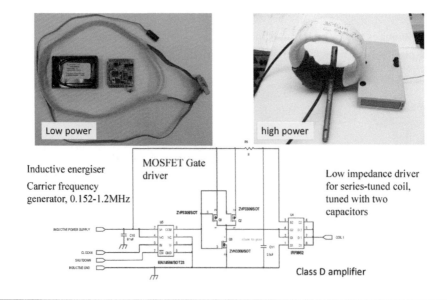

Low power

high power

Inductive energiser

Carrier frequency
generator, 0.152-1.2MHz

MOSFET Gate
driver

Low impedance driver
for series-tuned coil,
tuned with two
capacitors

Class D amplifier

15 Passive signalling by impedance modulation of implant coil

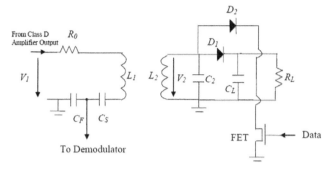

(a)

Equivalent circuit

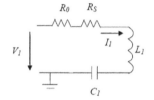

16 Link modulation

Link modulation

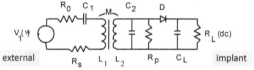

external   implant

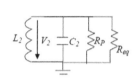

A useful property of coupled tuned circuits is that the secondary impedance Z_2 looking from the mutual inductance M will appear reflected in the primary circuit instead of M as series impedance $\omega^2 M^2 / Z_2$. Modulation of this reflected impedance by abrupt changes in R_2 enables data to be telemetered from the implant. Since the technique requires no separate power supply (or other hardware) it is known as 'passive signalling'.

Equivalent circuit

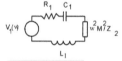

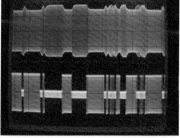

Primary current
50V/divn

Modulation
factor = 0.2

Secondary voltage
10V/divn

200us per divn

17 SCIMITAR block diagram

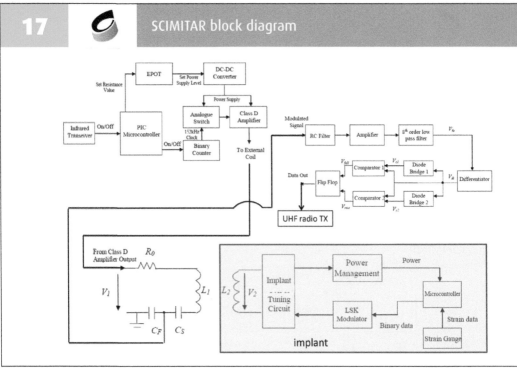

18 From implant forces to GUI display and data logging

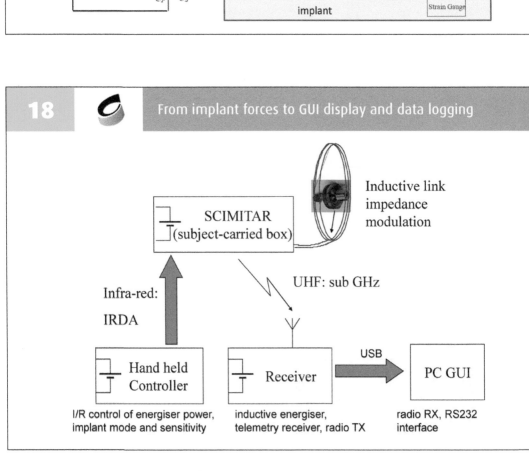

19

Inductive links used for power transfer and telemetry: power efficiencies

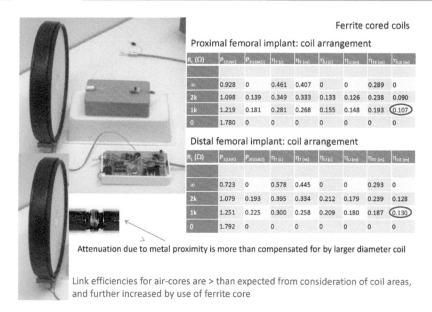

Ferrite cored coils

Proximal femoral implant: coil arrangement

R_i (Ω)	$P_{1(LINK)}$	$P_{2(LOAD)}$	$\eta_{T\,(c)}$	$\eta_{T\,(m)}$	$\eta_{U\,(c)}$	$\eta_{U\,(m)}$	$\eta_{TE\,(m)}$	$\eta_{UE\,(m)}$
∞	0.928	0	0.461	0.407	0	0	0.289	0
2k	1.098	0.139	0.349	0.333	0.133	0.126	0.238	0.090
1k	1.219	0.181	0.281	0.268	0.155	0.148	0.193	(0.107)
0	1.780	0	0	0	0	0	0	0

Distal femoral implant: coil arrangement

R_i (Ω)	$P_{1(LINK)}$	$P_{2(LOAD)}$	$\eta_{T\,(c)}$	$\eta_{T\,(m)}$	$\eta_{U\,(c)}$	$\eta_{U\,(m)}$	$\eta_{TE\,(m)}$	$\eta_{UE\,(m)}$
∞	0.723	0	0.578	0.445	0	0	0.293	0
2k	1.079	0.193	0.395	0.334	0.212	0.179	0.239	0.128
1k	1.251	0.225	0.300	0.258	0.209	0.180	0.187	(0.130)
0	1.792	0	0	0	0	0	0	0

Attenuation due to metal proximity is more than compensated for by larger diameter coil

Link efficiencies for air-cores are > than expected from consideration of coil areas, and further increased by use of ferrite core

---

20

Tibial nail soft encapsulated with adhesive medical grade silicone rubber

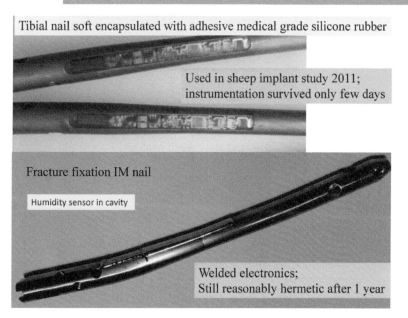

Tibial nail soft encapsulated with adhesive medical grade silicone rubber

Used in sheep implant study 2011; instrumentation survived only few days

Fracture fixation IM nail

Humidity sensor in cavity

Welded electronics; Still reasonably hermetic after 1 year

21 Implant transduction chain for femoral implants: uses analog electronics only

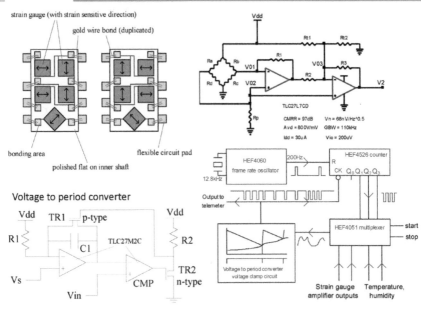

strain gauge (with strain sensitive direction)

gold wire bond (duplicated)

bonding area

polished flat on inner shaft

flexible circuit pad

TLC27L7CD

CMRR = 97dB   Vn = 68n V/Hz^0.5
Avd = 800V/mV   GBW = 110kHz
Idd = 30uA   Vio = 200uV

HEF4060 frame rate oscillator

HEF4526 counter

HEF4051 multiplexer — start / stop

Voltage to period converter voltage clamp circuit

Output to telemeter

12.8kHz   200Hz

Strain gauge amplifier outputs   Temperature, humidity

Voltage to period converter

Vdd   TR1   p-type

R1   C1   TLC27M2C

Vs   +   CMP   TR2   n-type

Vin

Vdd   R2

22 Tibial nail implant circuit: 4 strain channels

Tibial nail implant circuit: 4 strain channels, ADuC7061 (8 kSPS, 24-bit ΣΔ ADC, 16-bit/ 32-bit ARM7TDMI® MCU, and Flash/EE)

Fabricated on FR4 strip

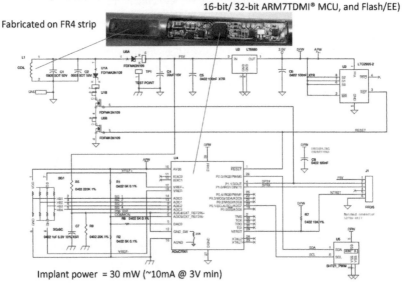

Implant power = 30 mW (~10mA @ 3V min)

23 Strain gauge characteristics

Strain gauge characteristics

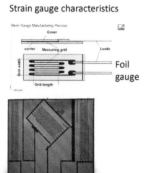

 Foil gauge

Gauge type	Strain sensitivity (gauge factor)	temperature coefficient (ppm/°C)	Unstrained drift rate
metal foil	2.1	Thermally match gauge to substrate (Measurements Group 1983)	20µε/year (with practice) (Freynik 1976)
thin film*	1.9	20	<0.02% of full scale per year
thick film	2 to 50 (Cattaneo 1980)	10 to 400 (Cattaneo 1980)	Approx. 2µε/year
semiconductor**	Up to 200	Up to 10,000	
vibrating beam***	700	3 (deduced)	0.03µε per 1000h at 50°C (deduced)

* Strain Measurement Devices Ltd, Sharon Road, Bury St Edmunds IP33 3TZ.
** Kulite Sensors Ltd, Marbaix House, Besemer road, Basingstoke RG21 3LG.
*** Haroda et al. 1986.

Thin film gauges

thick film circuit

Semiconductor strain gauge

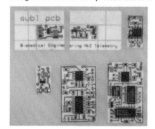

24 Implant electronics fabrication types

Implant electronics fabrication types

Analog SMD on thick film printed alumina

Analog and digital SMD on multi-layer flexible polyimide

Analog and digital SMD on multi-layer FR4

Electron beam welding

Sectioned welds

25 Instrumented implants incorporating ΣΔ ADCs and uC

Reverse shoulder: rotator cuff arthropathy - to understand the modes of failure, leading to enhanced glenoid fixation. Improved placement of the prosthesis in the bone.

 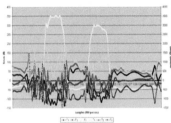 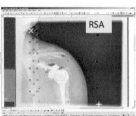

Resisted abduction at 6 months post-op

SmartNail: fracture fixation temporary nail, to indicate optimum removal time by strain redistribution between nail and bone.

26 Calibration options for instrumented implants

Techniques depend on:

- degrees of freedom measured
- how controllable is each d.o.f.
- load application: deadweight or semi-automatic (calibration rig)

Single linear regression usually sufficient

Multiple linear regression required as cannot control off-axis loading

1 d.o.f. deadweight loading

LabView control of 3 axes (force and torque x3) via stepper motors, with force feedback via 3 load cells. Forces to 1.4kN; torques to +-5Nm

27

Forces and moments from instrumented bone tumour replacement implants

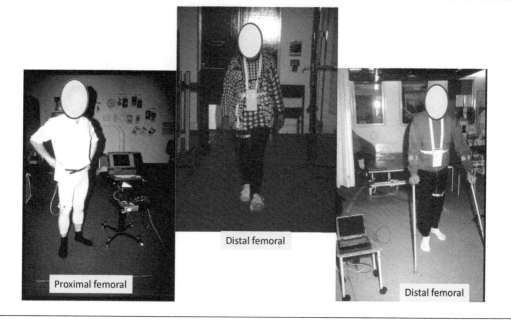

Proximal femoral

Distal femoral

Distal femoral

28

To study the effect of bone-replacing implants on the fixation

PFR fixation with offset loading

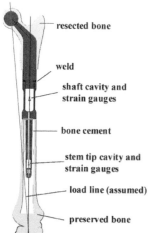

- resected bone
- weld
- shaft cavity and strain gauges
- bone cement
- stem tip cavity and strain gauges
- load line (assumed)
- preserved bone

4 patients;
2 year study each

Mk2 DFR: measurements

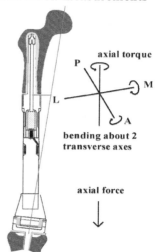

axial torque

P

L

M

A

bending about 2
transverse axes

axial force

29 Chondrosarcoma of proximal femur

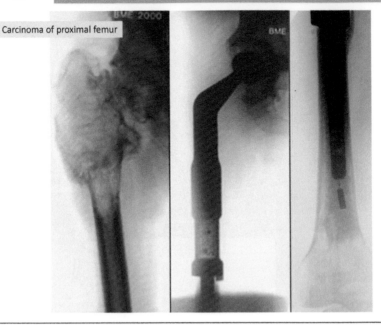

Carcinoma of proximal femur

30 Forces during walking cycle – Proximal femoral replacement

Axial force developed at shaft
And stem tip during gait

Proximal femoral replacement
At 6, 12, 18 months post-op

Gradual increase of tip / shaft ratio
Indicates progressive loosening:

Cannot be reliably quantified by
Radiographic observation

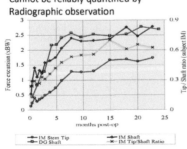

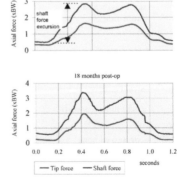

31 Force and moments (walking at 1 year)

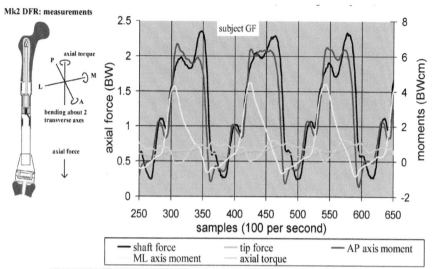

Forces and moments telemetered from two distal femoral replacements during various activities. J Biomech **34**, 2001. Taylor, Walker.

32 Typical gait cycle (compressed) for various activities at 1 year (subject GF)

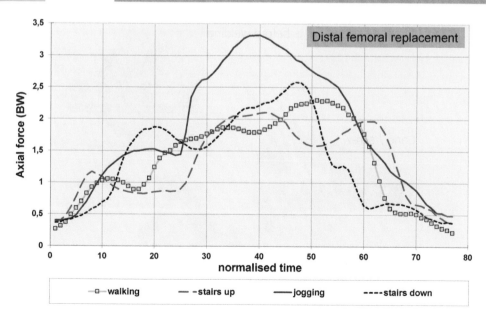

33 Distal femoral replacement. GF walking: Various activities

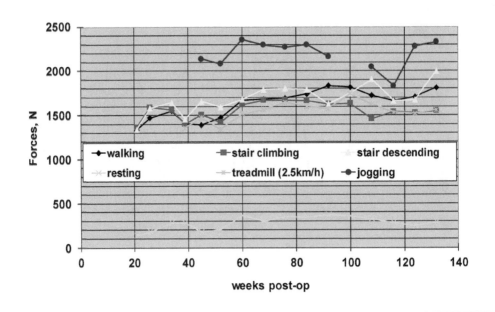

34 Instrumented reverse anatomy shoulder replacement

Glenoid implant components before welding

Cone / screw

Head / diaphragm

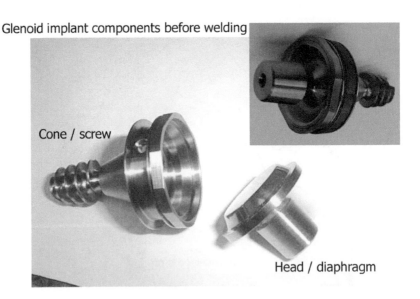

35 Intra operative

Intra operative

Glenoid component

Op date: 29 January 2009

36 Post operative RSA

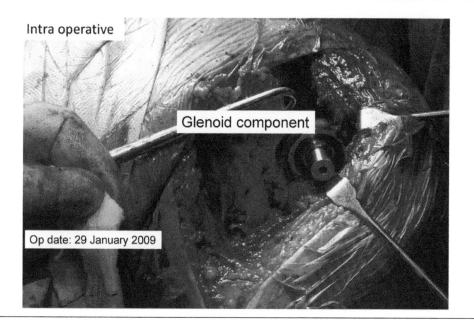

Post operative RSA

0.5mm tantalum balls welded to the implant for setting the calibration angle; also injected into the bone for migration measurements

37

Components of force and moments telemetered from reverse anatomy glenoid implant

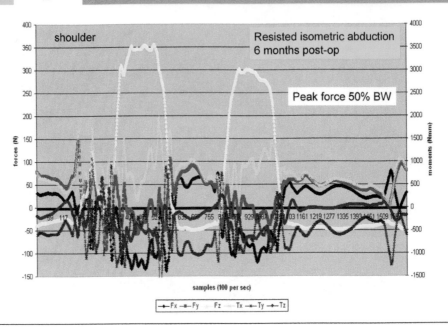

shoulder

Resisted isometric abduction
6 months post-op

Peak force 50% BW

samples (100 per sec)

Fx — Fy — Fz — Tx — Ty — Tz

38

Instrumented tibial tray for separate compartment knee force measurement

Instrumented tibial tray for separate compartment knee force measurement

±UCL

Strain gauge locations determined by FEA

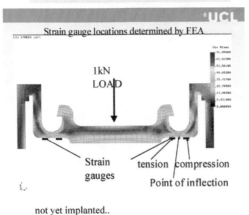

1kN
LOAD

Strain
gauges

tension compression

Point of inflection

not yet implanted..

separate compartments

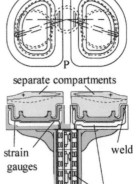

strain
gauges

weld

circuitry

silicone
rubber

gold wire coil

39 SmartFix: Sensing/distracting hexapod for optimum fracture healing

Intelligent frame:
Programmed prescription data
for many distraction
increments/day;
Regular bending stiffness meas.
using instrumented half pins

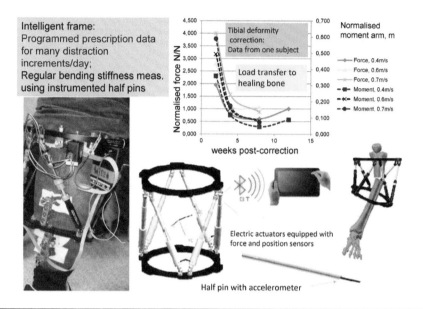

 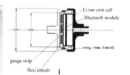

Electric actuators equipped with
force and position sensors

Half pin with accelerometer

40 Sensewheel

Sensewheel: a lightweight wheelchair wheel with embedded load cells and accelerometers to measure propulsion efficiency, thereby gathering data to refine the individual's rehabilitation programme using movement metrics. (Symonds, Taylor)

Many users **suffer** shoulder pain and injury in the long term because of unconscious overuse or poor technique.

Forces acting between the pushrim and the drivewheel are sensed by 3 strain gauged load cells. Data are digitised and transmitted to a separate data logger for processing and data display.

Health and Social Innovators Award 2014
Movement Metrics CIC formed, Jan 2015

41 The process of designing

The process of designing an instrumented implant, constructing and testing it to ensure safety both for the implant and the human subject in service, and carrying out measurements, are time consuming and expensive. It is perhaps not surprising therefore that this technology has not yet been scaled up for routine monitoring of implants, and is still largely a biomechanical research tool with very few instrumented subjects.

'Smart' implants are however now on the horizon.

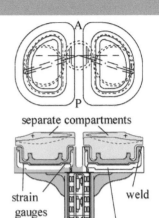

separate compartments

strain gauges

weld

circuitry

silicone rubber

gold wire coil

42 Acknowledgements

With thanks to all the students who have made contributions over the years.

CHAPTER 06

Implantable and Wearable Microelectronic Devices for Rebuilding the Lives of Disabled People

Hadi Heidari

University of Glasgow

In recent years, an enormous surge of works is being carried out on developing new technologies for microelectronics and sensing devices of various kinds of applications in the peripheral and central nervous systems. The miniaturised and implantable neural interfacing devices are widely applied in different clinical scenarios for example, peripheral, and spinal nerve interfaces for monitoring epilepsy, cochlear and retinal implants, and as deep brain stimulators thanks to evolutions in microelectronics, sensors, communications, and neuroscience. Implementation of such neuranics (neural electronics) using traditional CMOS technology is inefficient in terms of size, performance and power. Such inefficiencies have driven a significant effort to investigate the development of beyond-CMOS devices.

Implementation of microelectronics with innovative technologies such as spintronics and biocompatible materials provides an opportunity to produce and optimise powerful neuranics offering new diagnostics and therapies for more complex diseases and disabilities. The seminar will introduce our research at the Microelectronics Lab (meLAB) in the field of implantable and wearable microelectronics as well as magnetic sensors addressing the development of architectures for the implementation of efficient implantable neural probes and magnetomyography. It presents developments and trends in the neural recording devices including materials, sensors and circuit Interfaces focusing on the development of various types of implantable devices for the repair and diagnose of the diseased brain and movement disorders. Various approaches for obtaining highly sensitive magnetic sensors for implantable magnetomyography will be presented. Furthermore, the seminar will be concluded by presenting our research on flexible sensors to implement wearable intelligent wristworn for gesture recognition comprehensively.

1 Outline

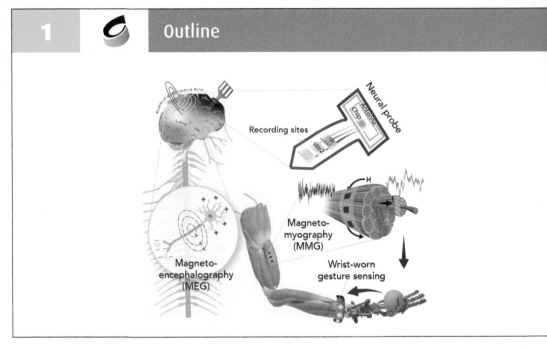

This presentation will cover our research at the University of Glasgow in three domains: (1) Wearables: wrist-worn and eyeball gesture control; (2) implantables: magnetomyography and ne probes; and (3) energy harvesting for wearable implantables.

2 Neurotechnology devices

Hybrid Enhanced Regenerative Medicine System (HERMES)

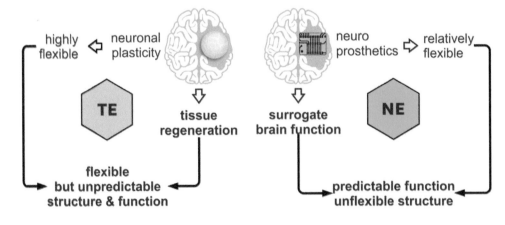

€8.4 M for brain repair using enhanced regenerative medicine

3 ## Neural interfaces

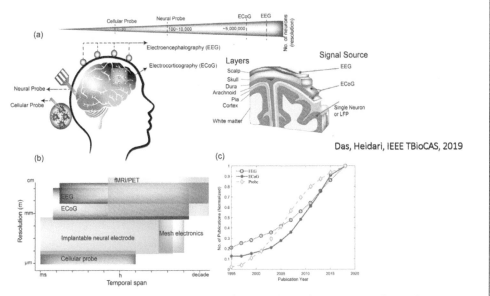

Das, Heidari, IEEE TBioCAS, 2019

(a) Neural recording systems. (b) Principles and physical constraints of neural recording technologies. (c) current research trend on neural interfaces

4 ## Key parameters (I)

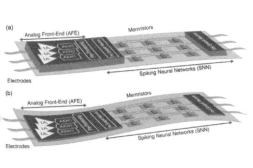

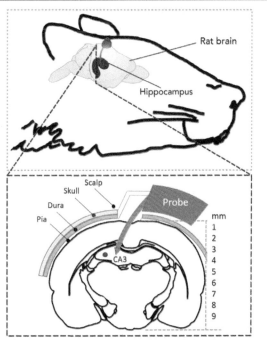

Integration of CMOS and memristors on flexible substrate on a neural probe

flexible neural probe in rat brain

5 Key parameters (II)

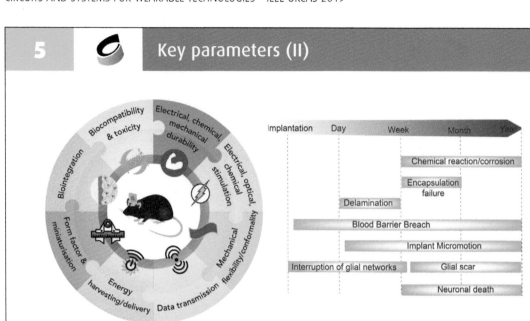

Key parameters in neuroscience research

Failure modes of neural device

6 Biointegration (I)

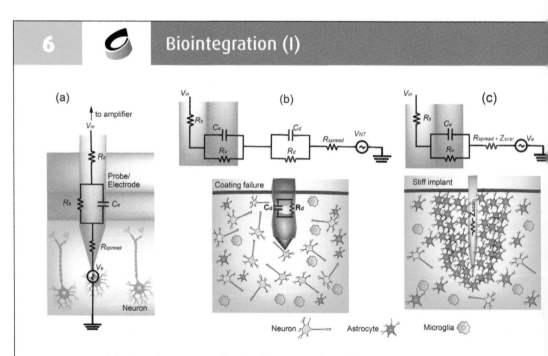

(a) Equivalent circuit of probe/tissue interface (b) Insulation damage creates undesired current pathways (c) Influence of neuroinflammatory reaction

7 Biointegration (II)

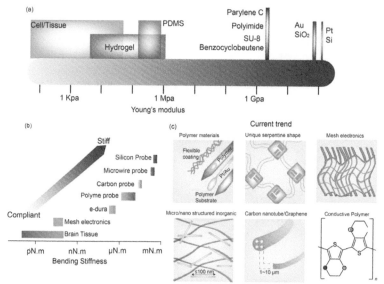

(a) Young's modulus of tissue & materials (b) Bending stiffness among implantable neural devices. (c) Current trend in implantable neural devices

8 Implantable magnetic sensors (I)

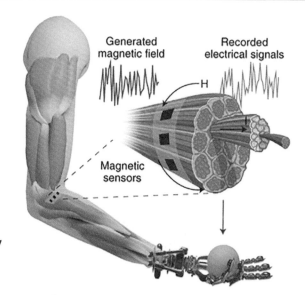

Magnetomyography (MMG)

Movement Disorders and Human-machine interface

9 Implantable magnetic sensors (II)

MMG sensor
No direct contact

EMG sensor
Myoelectric electrodes with direct contact

I H

Magnetic field lines

Muscle impulse

Muscle fibre

Zuo, Heidari, et al. IEEE Electron Device Letters, 2018

MMG vs. EMG

Tunneling Magnetoresistive (TMR) Sensors

Free layer
Barrier (MgO)
Reference
Antiferromagnetic

Proposed TMR

Oxide
Via
Metal layers
Gate
Bulk Source Drain
Si Substrate

Readout circuitry

10 Implantable magnetic sensors (III)

MMG Signal Modelling

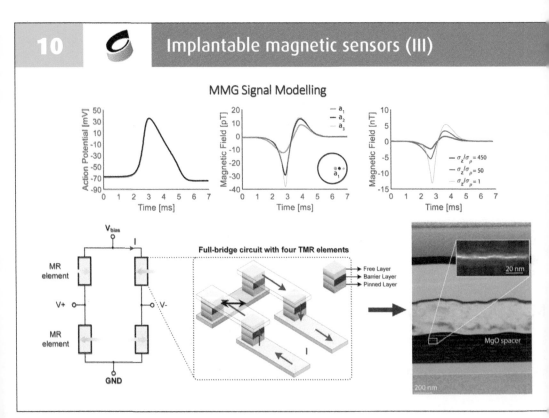

Full-bridge circuit with four TMR elements

V_{bias}
I
MR element
V+   V-
MR element
GND

Free Layer
Barrier Layer
Pinned Layer

20 nm
MgO spacer
200 nm

11   Implantable TMR sensors

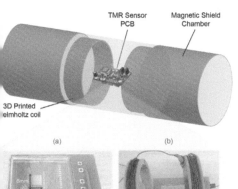

TMR Sensor PCB

Magnetic Shield Chamber

3D Printed Helmholtz coil

(a)   (b)

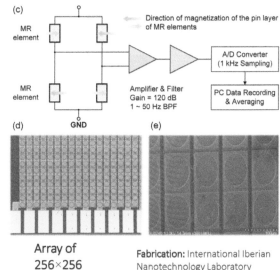

(c)

MR element

MR element

GND

Direction of magnetization of the pin layer of MR elements

Amplifier & Filter
Gain = 120 dB
1 ~ 50 Hz BPF

A/D Converter
(1 kHz Sampling)

PC Data Recording
& Averaging

(d)   (e)

Array of
256×256
Devices

Fabrication: International Iberian
Nanotechnology Laboratory

12   Implantable energy harvesting

Implantable Power harvesting

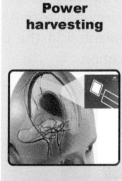

Independent system

Solar power   Body motion   Thermal

Self-powered

Transferring mechanism

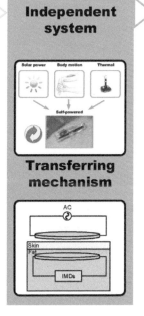

AC

Skin
Fat

IMDs

Solar cell

Piezoelectric

Biofuel

Thermalelectric

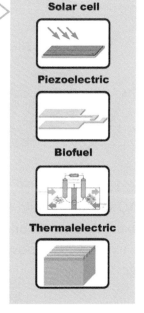

13 — Energy and power technologies

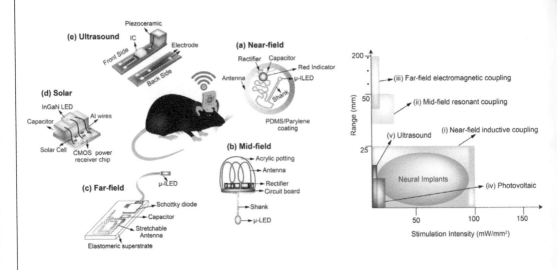

Various solutions for wireless technologies
in implantable neural probe

Comparison among wireless
technologies

14 — Implantable PV cells (I)

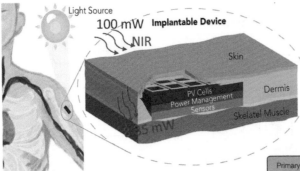

Implantable photovoltaic energy harvesting system implanted in the optimal location.

IEEE Access 2018
IEEE JERM 2019

PEG, TENG and WPT are considered as AC sources, while PV cell is a DC, relaxing the additional loss due to the extra AC-DC power conversion stage.

IEEE SENSORS 2018
IEEE ICECS 2018

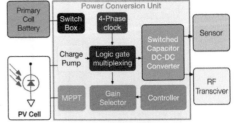

15 Implantable PV cells (II)

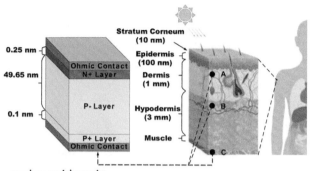

Stratum Corneum (10 nm)
Epidermis (100 nm)
Dermis (1 mm)
Hypodermis (3 mm)
Muscle

0.25 nm — Ohmic Contact, N+ Layer
49.65 nm — P- Layer
0.1 nm — P+ Layer, Ohmic Contact

Implantable photovoltaic energy harvester with is PPN back-surface illuminated structure.

A) under epidermis,

B) under dermis => light scattering causes a 47% drop in power at 550 nm and 9% at 850 nm

C) under adipose => 28% drop using NIR light and 52% using white light

Our proposed implantation location (between B & C), illumination wavelength (NIR light) and optimised PV cell harvest **> 8.35 mW/cm²** power

IEEE JERM 2019

ompared with other technologies, PV cells can produce energy in the tens to hundreds of microwatts ge within a relatively small area (mm² scale).

Ultrasonic power harvesting is one way to overcome these drawbacks, but the amount of power is lower than 1 mW.

16 Implantable PV cells (III)

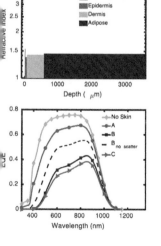

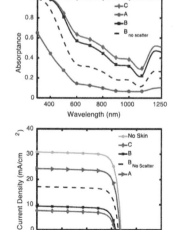

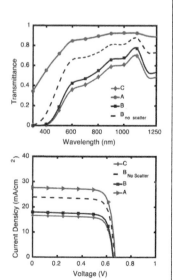

17 Wearable multi-sensor device (I)

Fusion of Wearable and Contactless Sensors for Intelligent Gesture Recognition

Advanced Intelligent Systems 2019
IEEE Sensors 2019

18 Wearable multi-sensor device (II)

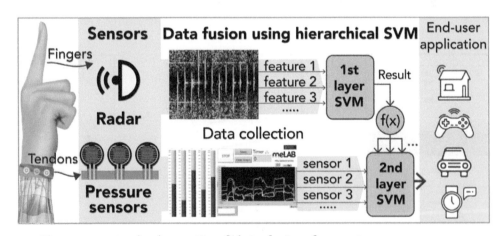

The conceptual schematic of data fusion for gesture sensing with Hierarchical SVM.

19 Wearable multi-sensor device (III)

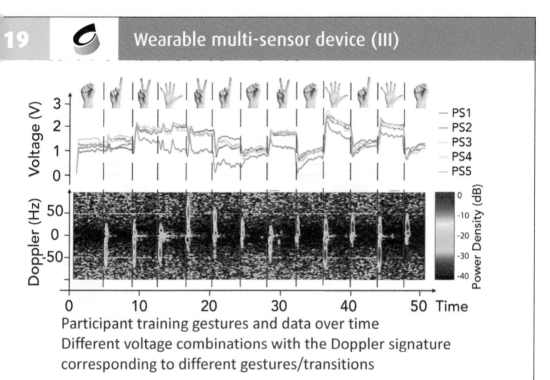

Participant training gestures and data over time
Different voltage combinations with the Doppler signature
corresponding to different gestures/transitions

rticipant 5's training gestures and data over time
s an example ('PS' in the figure denotes pressure
or). The participant was required to show a
ence of gestures and keep fingers still for four
nd for each. It can be seen that different voltage
binations are measured for different gestures,
here are also similarities that may make the
ification problem challenging. For the radar part,
Doppler signature corresponding to different

transition parts are illustrated, where the positive
Doppler corresponds to the movement approaching
towards the radar, and vice versa. The color indicates
the strength of the gesture movement with respect
to different parts of hand. The main misclassification
takes places between those classes for which the
directions of the movement are very similar, and the
amplitudes of the recorded signals are also similar
(e.g.1 to 5 and 2 to 5).

20 Wearable multi-sensor device (IV)

onfusion matrix of pressure sensors only. b)
 usion matrix of radar only. In confusion matrices,
diagonal elements are the events correctly
fied, whereas the non-diagonal elements are

the misclassifications. c) The result in first trial, both
cases of the highest accuracy and improvement are
tested. d) Confusion matrix after data fusion.

20 Wearable multi-sensor device (IV)

Confusion matrix of individual sensors before and after fusion

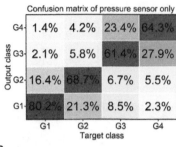
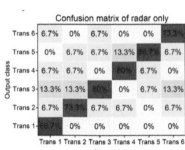
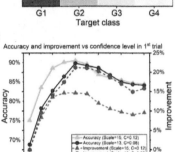

21 Wearable multi-sensor device (V)

Data Source	Object	Fusion method	Accuracy before fusion [%]	Final accuracy [%]	Improvement [%]	Computational burden	Reference
EMG + accelerometer	Gesture	HMM	85.5	91.7	6.2	Medium	ACM Conf. Intell. User Int. 2009
RGBD + Inertial	Human action	CNN+ LSTM	79.1	92.8	13.7	Intensive	IEEE Sensors Journal 2018
Optical sensor + depth camera + radar	Gesture	CNN	90.9	94.1	3.2	Intensive	IEEE Sensors Journal 2014
Radar + optical imagery	Land cover	Multi-SVM	68.9	77.1	8.2	Low	Inf. Fusion 2013
Inertial + radar	Human activity	SVM+ KNN	89.3	97.8	8.5	Intensive	IEEE JERM 2018
Pressure sensors + radar	Gesture	HSVM (linear)	69.0	92.5	23.5	Low	This work Adv. Intell. Syst. 2019

Fusion of Wearable and Contactless Sensors for Intelligent Gesture Recognition

22 ## Magnetoresistive device

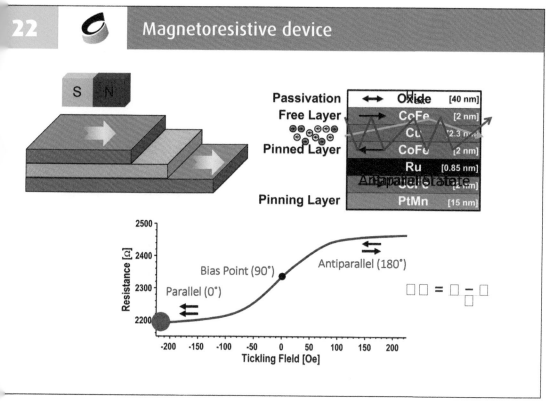

A spin-valve is an **elaborately engineered multilayer thinfilm** structure as shown here.

The device is composed of a **pinning layer** which sets the orientation of the synthetic antiferromagnet.

The synthetic antiferromagnet is composed of a CoFe/Ru/CoFe trilayer structure. The **magnetization** of the two layers are **pinned** in the **antiparallel** direction through RKKY coupling. The top CoFe layer is called the free layer and the magnetization **rotates freely** to align with the **external magnetic field**.

This structure is then **passivated** with a thin oxide to **isolate** it from the biological and chemical solutions.

The spin-valve exhibits a **quantum mechanical** effect known as **spin-dependent scattering** occurring primarily at the CoFe/Cu/CoFe interfaces (although the bulk scattering needs to be accounted for as well).

Let me **illustrate this concept** by examining the **two extreme cases**.

In the **parallel** configuration, the electrons pass through the device **without significant spin-**

dependent scattering because the two layers are **aligned**. The resistance in this state is the lowest.

- In the **anti-parallel** configuration, the electrons pass through the device and undergo significant spin-dependent scattering leading to a **higher resistance**, as seen here.

- The **quiescent state** is set by the shape of the sensor (or an external bias field).

- The **magnetoresistance ratio** (MR) for GMR SV is **typically 5-15%** depending on the materials used.

- I should point out that most conductors exhibit this phenomenon, just on such a minute scale that it is not useful for sensors (**PPM**).

Depending on the orientation of the pinned layer and the free layer, as electrons pass through this device,

Extremely sensitive, comparable to earth's magnetic field < 0.5 Oe (50 uT)

An external bias field (or shape anisotropy) sets the quiescent state of the CoFe free layer

23 — Translation – Magnetic biosensors for malaria

External Uniform Magnetic Field

Magnetized Paramagnetic Particle

Red Blood Cell

Parasite

Cling Film (PVC)

Sensor

Magnetoresistive Biosensor

SFC GCRF Grant, Glasgow Institute of Biodiversity Animal Health & Comparative Medicine with Uganda

Healthy RBC   Infected RBC (N = 50)   Infected RBC (N = 100)

Paramagnetic substances have a magnetic moment only in the presence of an applied magnetic field. Paramagnetism in hemozoin arises from the unpaired electrons of the Fe^{3+} species of iron.

24 — Malaria point-of-care diagnostics

(a) Slice: Magnetic field norm (A/m)
Arrow Volume: Magnetic flux density

(b) Slice: Magnetic flux density norm (T)
Arrow Volume: Magnetic flux density

Paramagnetic Particle

(c) Healthy Red Blood Cell

(d) Malaria Infected Red Blood Cell

(e) Magnetic Field [Oe] vs Magnetization [T]

(f) Output Voltage [µV] vs X-Axis Bead Displacement [µm]

(g) Output Voltage [mV] vs Number of Particles
— Malaria Infected RBC
— Healthy RBC

Magnetic sensor

Cartridge sample handler

15 cm

Touch screen LCD

25 Microelectronics Lab (meLAB)

www.melabresearch.com

3 Research Associates, **6** PhD Students

 University of Glasgow

 EPSRC
Engineering and Physical Sciences Research Council

 Scottish Funding Council

 国家自然科学
基金委员会
National Natural Science
Foundation of China
N S F C

 THE ROYAL SOCIETY

---

26 References

1. Heidari, H. (2018) Magnetoelectronics: Electronic skins with a global attraction. *Nature Electronics*, 1(11), pp. 578-579. (doi:10.1038/s41928-018-0165-2)

2. Liang, X., Li, H., Wang, W., Liu, Y., Ghannam, R. , Fioranelli, F. and Heidari, H. (2019) Fusion of wearable and contactless sensors for intelligent gesture recognition. *Advanced Intelligent Systems*, (doi:10.1002/aisy.201900088) (Early Online Publication)

3. Zuo, S., Nazarpour, K. and Heidari, H. (2018) Device modelling of MgO-barrier tunnelling magnetoresistors for hybrid spintronic-CMOS. *IEEE Electron Device Letters*, 39(11), pp. 1784-1787. (doi:10.1109/LED.2018.2870731)

4. Fan, H., Li, D., Kelin, Z., Cen, Y., Feng, Q., Qiao, F. and Heidari, H. (2018) A 4-channel 12-bit high-voltage radiation-hardened digital-to-analog converter for low orbit satellite applications. *IEEE Transactions on Circuits and Systems I: Regular Papers*, 65(11), pp. 3698-3706. (doi:10.1109/TCSI.2018.2856851)

5. Heidari, H. , Bonizzoni, E., Gatti, U. and Maloberti, F. (2015) A CMOS current-mode magnetic hall sensor with integrated front-end. *IEEE Transactions on Circuits and Systems I: Regular Papers*, 11(4), pp. 1270-1278. (doi:10.1109/TCSI.2015.2415173)

6. Liang, X., Ghannam, R. and Heidari, H. (2019) Wrist-worn gesture sensing with wearable intelligence. *IEEE Sensors Journal*, 19(3), pp. 1082-1090. (doi:10.1109/JSEN.2018.2880194)

7. Nabaei, V. , Chandrawati, R. and Heidari, H. (2018) *Magnetic biosensors: modelling and simulation. Biosensors and Bioelectronics*, 103, pp. 69-86. (doi:10.1016/j.bios.2017.12.023) (PMID:29278815)

8. Lei, K.-M., Heidari, H. , Mak, P.-I., Law, M.-K., Maloberti, F. and Martins, R. P. (2017) A handheld high-sensitivity micro-NMR CMOS platform with B-field stabilization for multi-type biological/chemical assays. *IEEE Journal of Solid-State Circuits*, 52(1), pp. 284-297. (doi:10.1109/JSSC.2016.2591551)

CHAPTER 07

Development of Nanowire Based Sensors for Integration with Wearable Systems for Real Time, Continuous and Non-Invasive Health Monitoring

Ian Sandall

University of Liverpool

Wearable sensors are "smart" electronic devices that can be worn on the body, or even implanted to measure a range of data. Nearly all commercially available wearable devices, are focused on the personal health and fitness market, typically recording physical characteristics; such as number of steps taken, heart rate and calories burnt. Recent advances in Microelectronics, telecommunications, and sensor manufacture are opening up the possibility to develop semiconductor based wearable sensors, capable of providing actual real-time, continuous monitoring for a range of illnesses and medical conditions.

The detection of bio markers within bodily fluid can be used to monitor a range of conditions, the most commonly used fluid for such diagnosis is blood, however this is not compatible with continuous, simple to use wearable systems. Over recent years a range of other bodily fluids have been investigated as alternatives including; Interstitial Fluid, Sweat, Tears and Saliva. Sweat presents itself as an ideal candidate for wearable based sensors, where the system can sit on the skin and absorb sweat to perform diagnosis.

Semiconducting nanowires offer the potential to act as an efficient and low cost-based platform for such sensing. However current fabrication methods of such devices are often time consuming and expensive, limiting their development. This is further hampered by many of these nanowires being developed on non-optimum substrates for integration with a wearable platform.

In this talk, we will explore the use of dielectrophoresis as a manufacturing approach for a nanowire-based sensor platform. The exploitation of dielectrophoresis enables nanowire based electronic sensors to be rapidly and efficiently fabricated with a high degree of uniformity and repeatability. Furthermore, the use of this technique allows the nanowires to be transferred to other substrates, with this in mind we will detail an approach to realize a paper-based sensor. The resultant sensor will be evaluated by looking at its sensitivity to ionic solutions as a first step to developing a sweat based sensor.

The talk will conclude by looking at the remaining challenges for this technology, including functionalizing of the nanowires to provide selective screening and discussing the requirements for associated control and readout electronics.

1   Wearable sensors (I)

- Sensors that can be worn on the body

- Most common example today is fit-bit type devices
  - Sense Heartbeat
  - Movement
  - etc

- Can more "medical based" wearables be achieved?

Wearable biosensors are smart electronic devices that can be worn on the body as implants or accessories. Recent advances in microelectronics, telecommunications, and sensor manufacturing have opened up possibilities for using wearable biosensors to continuously monitor an individual's body status without interrupting or limiting user's motions. Nearly all commercially avail wearable electronics focus on tracking users' phy activities, devices that can provide an insightful v of user's health status at the molecular level rec considerable development.

2   Wearable sensors (II)

Although blood is by far the most understood sample for diagnosis, it poses significant challenges and risks in extracting it for sensing purposes. Other biological fluids such as Interstitial Fluid, sweat, tears, and saliva also contain tremendous biochemical analytes that can provide

- For medical / biological based sensing the most commonly used fluid is blood

- Better alternatives for wearable sensing include:
  - Interstitial Fluid
  - Sweat
  - Tears
  - Saliva

- Open possibility for non-invasive continuous monitoring

[1] Kim et al., Acc. Chem. Res. 2018, 51, 2820–2828
[2] Kim et al., Adv. Sci. 2018, 5, 1800880

valuable information and are much more readily accessible compared to blood. However their use is less well developed and careful consideration and calibration is required to obtain reliable and trusted information from these fluids.References:

[1] Kim et al., "Wearable Bioelectronics: Enzyme-Based Worn Electronic Devices", Acc. Chem. Research, 51, pp 2820 (2018)

[2] Kim et al. "Simultaneous Monitoring of Sweat and Inte Fluid Using a Single Wearable Biosensor Platform", Adv. Sci., 1800880 (2018)

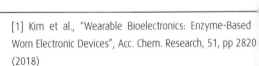

3 Wearable sensors – Key problems

- What can be used as the sensor?

- How will bodily fluid be introduced / controlled?

- How will user receive / understand the resultant data?

ere are 3 key questions that need to be addressed ⟩ achieve such a wearable sensor system. 1) What the sensing element be, what form of electrical/ ⌐ical sensor, it needs to be bio-compatabile, light weight and low cost. 2) How will the body fluid of interest be placed onto the sensor and 3) How will the system be controlled / provide a readout to the user. This work and talk focuses on questions 1 and 2.

4 What can be used for sensing bodily fluids?

- Semiconducting Nanowires.....

<100 nm

Typically 1 – 100 um

[3] Ambhorkar, et al. Micromachines, 9, 679 (2018)
[4] Liu et al., ACS Nano 2018, 12, 1170–1178

4 What can be used for sensing bodily fluids?

Semiconducting nanowires, offer the potential to act as the sensing element and answer question 1 from the previous slide. These "wires" sometimes called nano-pillars or nano-whiskers are thin long pieces of semiconductor. By definition they have a diameter of under 100nm, they have typical lengths of a few to a few hundred microns. The sketch shows the nanowire having a cylindrical shape, it is more typical for it to have a hexagonal cross section. The Electron microscope images show random growth of nanowires with a distribution of diameters, heights and alignments. This is typically how they are formed. Given their small size they are light and

able to interact with biological molecules for sens They are typically grown on large substrates (2 – in diameter) providing high density of nanowire, sensor only requires a few (10 – 1000) nanowires such the cost is virtually zero. Therefore nanow can provide light, versatile and low cost element

[3] Ambhorkar et al. "Nanowire-Based Biosensors: From Grow Applications", Micromachines, 9 pp 679 (2018)

[4]4 Liu et al., "Highly Sensitive and Wearable In203 Nanorib Transistor Biosensors with Integrated On-Chip Gate for Glu Monitoring in Body Fluids", ACS Nano 12, pp 1170-1178 (201

5 Why consider nanowires

- **Due to high ratio of surface area to volume nanowires are very sensitive to their environment**

- **Arsenic and Antimony based nanowires are especially sensitive due to high carrier mobilities and typically narrower bandgaps**

- **In(Sb)As Nanowires are therefore attractive candidates for biological and chemical sensing by monitoring changes in surface charge**

- **NWs by definition are small for integration into a wearable system**

Nanowires are attractive as potential sensors as they have high area / volume. This makes them very sensitive to the electrical properties of their surrounding medium. Most nanowire work

has focused on either Silicon or Carbon Nanotu however III-V semiconductor based nanowires, h much higher carrier mobilites and as such be m sensitive to their environments.

6 ## Issues with nanowires (I)

- While Nanowires can be grown on a range of substrates (i.e. Silicon, III-V, Ge and even glass) enabling low cost and high volume production. None of these are flexible for use in wearables and some pose toxicity risks.

- How can they be integrated to "sense" bodily fluids?

rrent growth methods for nanowires use rigid and potentially toxic substrates, these will need to be ntegrated into a wearable platform.

7 ## Issues with nanowires (II)

- The predominate method to fabricate nanowires into devices, does not lead itself to easy use / integration as a sensor.

- Vertical Devices

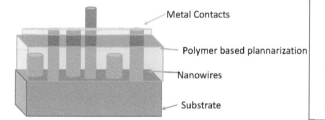

Metal Contacts

Polymer based plannarization

Nanowires

Substrate

Potential Issues

- Not all NWs contacted

- Some NWs provide potential short

- Final device performance will depend on all properties of all contacted NWs

- Surface of majority of NWs are heavily Screened by planarization layer

e majority of nanowire devices realized to te, are so-called vertical devices, where the wires as planarized and passivated. This is good Ds. Large bulk diodes and transistors, but not for sensing as the surface of the nanowires are shielded, also this technique utilizes all the nanowires meaning overall performance can be dominated by 1 or 2 poor quality wires.

8		Issues with nanowires (III)

Nanowires can be transferred onto other substrates for better integration in flexible/wearable approach and to change geometry to improve sensing.

➤ Horizontal Devices – Drop Casting Method

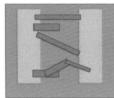

Potential Issues

- Not all NWs connected

- No guarantee regarding the quality of the selected NWs.

- Complicated NW-NW junctions

- Random Orientation

- Final device performance will depend on all properties of all contacted NWs

So called horizontal devices can be realized, by removing the nanowires from their native substrate and drop casting on a new one. This new substrate could be flexible plastic or a paper based system for integration in wearable technology. However the nanowire alignment will be random relative to the electrical contacts so performanc unrepeatable and difficult to model, due to mul nanowire-nanowire junctions. As with ver devices performance can again be dominated handful of poor nanowires.

9		Problems with nanowires

- Alternative approaches can give better alignment for device fabrication and sensing, but at much higher cost

- Horizontal Devices – Pick & Place

Potential Issues

- Slow and relatively expensive process

- No guarantee regarding the quality of the selected NWs.

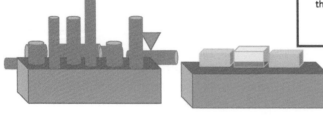

9 Problems with nanowires

n alternative approach to horizontal devices uses an AFM to "pick and place" individual nanowires, vever this is a slow and expensive process so not able for low cost sensor manufacture. Also there is no guarantee "good" nanowires will be moved so the process has a low yield and is difficult to reproduce.

10 Utilizing dielectrophoresis as another option

➤ During a Dielectrophoresis Process (DEP), particles as dispersed in solution and subjected to an AC Electric field

Charge distribution on the particles will act as rotating diploe, enabling the particle to line up to the field

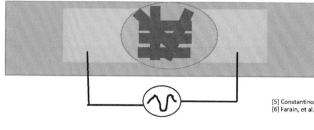

Can use anything for the substrate (i.e. glass, flexible plastic, paper)

[5] Constantinou et al., ACS Nano, 2016 10 4384-4394
[6] Farain, et al. APL, 113, 063101 (2018)

e process of dielectrophoresis may offer an ttractive alternative production process, in which nanowires can be drop-cast onto a new substrate ' cost) but the process will ensure alignment with trical contacts and repeatablity. Dielectrophersis ks by applying an alternating electric field over 2 electrical contacts, when the nanowires are ersed into this gap, charge is accumulated on the of the nanowires, this charge will oscillate with the frequency of the field causing the nanowires to align across the electrode gap.

[5] Constantinou et al., "Simultaneous Tunable Selection and Self-Assembly of Si Nanowires from Heterogeneous Feedstock", ACS Nano, 10, pp. 4384-4394 (2016)

[6] Farain, et al., "Universal rotation of nanowires in static uniform electric fields in viscous dielectric liquids", Applied Physics Letters, 113, 063101 (2018)

11 Utilizing dielectrophoresis as a 3rd option

Potential Advantages if using this approach

➢ Nanowires are selected by their length so none are "lost"

➢ Simpler and more cost effective than TEM/AFM based pick and place

➢ Nanowires will be deposited at known locations

➢ Simple to passivate and integrate microfluidics over the nanowire

➢ As the AC field frequency increases only highly conductive nanowires are likely to be able to react to the field – potentially leading to higher quality nanowires and devices being realized

This approach has a number of potential advantages

12 DEP setup (I)

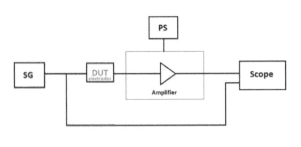

A typical experimental setup for dielectrophoresis, showing a glass slide based substrate. A signal generator provides the AC signal between the electrodes the resultant current between electrodes is monitored via a transimpeda amplifier and oscilloscope.

13 **DEP setup (II)**

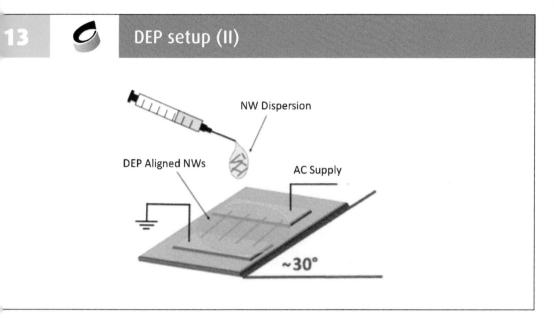

hen conducting the technique, the substrate is often held at a slight tilt to facilite run-off from the solvent and aid evaporation.

14 **DEP setup (III)**

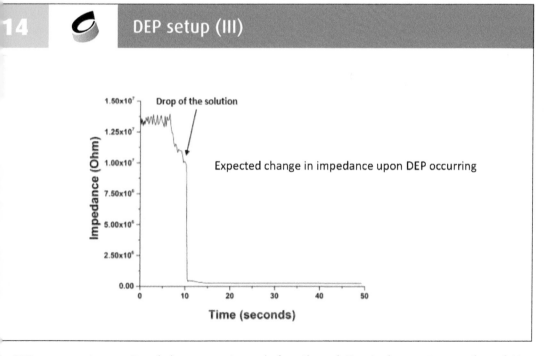

DEP process is monitored by measuring e impedance between the electrode gap function of time. This is done by using a impedance amplifier and measuring the current across the gap as shown several slides ago. In xample shown the impedance is measured from before the solution is drop cast, once the solution is dispersed (at approximately 8 seconds) there is a sharp drop in the measured resistance. This is caused by the nanowires aligning across the electrode gap and providing current flow.

15 Nanowire growth and fabrication

➢Nominally undoped InAs Nanowires grown by random seed MBE on Silicon Substrates

➢ Random distribution of Nanowires with Lengths in the range 2 – 5 µm, average size of 3.4 µm

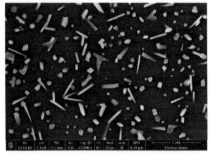

The nanowires used in this work, were Indium Arsenide (InAs) nanowires and were grown via Molecular Beam Epitaxy at Lancaster University. The growth gives a random distribution of nano lengths, these were then removed from the subs via sonication and dispersed in solvents.

16 Device fabrication

➢ Nanowires dispersed into Deionized Water via sonication and drop cast onto substrates with interdigitated contacts (Al and Au) to form asymmetric contacts.

Initial testing was done on glass substrates patterned with interdigit electrodes of Aluminium and Gold, with a varying electrode separation from 1 – 10 microns, enabling different sized nanowir be utilized in the same device.

17

Influence of AC frequency during DEP (I)

e force experience
y the nanowires due
ne electric field, will
end on the electrical
perties (conductivity
permittivity) of
nanowires and
medium they are
ersed in, as well as
applied frequency.
force can be
ulated from the
sis Mossotti factor
shown, as can be
n as the frequency
eases the resultant
e decreases, as

➤ To look at the influence of the applied AC frequency on the nanowires we have calculated the force exerted on the NWs as a function of the applied frequency, using the Clausis Mossotti Factor

$$F_{DEP} = 2\pi\varepsilon_0\varepsilon_m a^3 \frac{\omega^2\left(\varepsilon_p - \varepsilon_m\right)\left(\varepsilon_p + 2\varepsilon_m\right) + (\sigma_p - \sigma_m)(\sigma_p + 2\sigma_m)}{\omega^2\left(\varepsilon_p + 2\varepsilon_m\right)^2 + (\sigma_p + 2\sigma_m)^2}$$

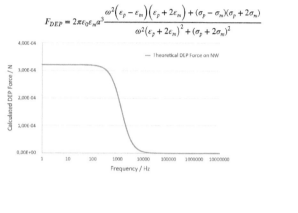

n only high quality, defect free nanowires, with a high conductivity will feel a significant force at these uencies.

18

Influence of AC frequency during DEP (II)

om the experimental
neasurement of
impedance a
acteristic time
tant for the
process can be
rmined from the
over which the
edance decreases.
time constant
ld be inversley
ortional to the force
e previous slide, i.e.
bigger the force the
ter time needed for
ıment.

➤ During the deposition of the NWs via DEP we have measured the time constant of the impedance signal at varying applied frequencies and have compared this to the force plot

$$\tau = \frac{-t}{\ln(\frac{Z(t) - Z_0}{Z_{max}})}$$

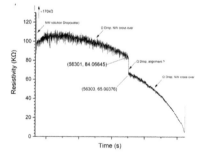

19   Influence of AC frequency during DEP (III)

➢ During the deposition of the NWs via DEP we have measured the time constant of the impedance signal at varying applied frequencies and have compared this to the force plot

$$\tau = \frac{-t}{\ln(\dfrac{Z(t) - Z_0}{Z_{max}})}$$

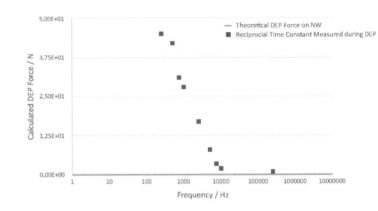

The measured time constants (red squares) give good agreement to the predicted forces as we sw across frequency

20   Influence of AC frequency during DEP (IV)

➢ After the solvent had fully evaporated the devices were tested as Schottky diodes

➢ For devices fabricated with a slow DEP frequency a near symmetric IV shape was obtained

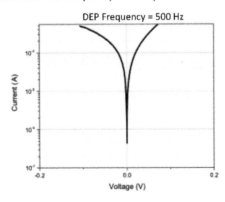

Once the devices were fabricated we tested them as schottky diodes to characterize performance. The image shows the resultant current-voltage (IV) characteristic for a device fabricated at a slow DEP frequency of 500 Hz. The IV curve shows symmetry indicating poor diode performance suggesting that the nanowires were just acting highly conductive path across the device.

21 Influence of AC frequency during DEP (V)

➤ As the AC frequency is increased more diode like characteristics are obtained, suggesting only "high quality" nanowires are captured during these DEP processes

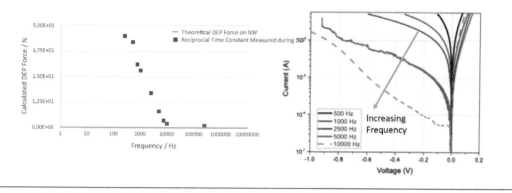

e have evaluated the IV's for devices fabricated at different DEP frequencies as the frequency increases the diodes become more metric and rectifying in nature. This indicates that at higher DEP frequencies only the "better nanowires" are influenced by the DEP force and positioned across the electrodes, resulting in much improved diode performance.

22 How can we use these in a wearable platform

• Work so far has been on quartz substrates

• In principle can be small enough to be incorporated into a wearable chip

• However this would require additional microfluidics to select and move fluids

> Possible Alternative – Paper based sensing

hile this is encouraging in terms of fabricating devices with nanowires, further work is needed velop this into a sensor platform, furthermore the use of quartz glass is non ideal for a wearable platform. An alternative may be to make use of paper substrates

23 Utilizing paper substrates

- Can nanowires be deposited (and contacted) on appropriate filter paper

- This can then be placed on skin to absorb fluid (i.e. sweat), which in turn contacts the nanowires and produces a measurable electric change

- Could be used as a "disposable" sensing element in a wearable system

We have investigated using filter papers as substrates and depositing the nanowires onto these. This could then be used a disposable sensing element within a wearable system as both the paper and relative number of nanowires nee would be very low cost.

24 Fabricating into a proof of concept sensor

To test this a proof of concept device was realized, incorporating printed contacts onto a piece of filter paper, a similar interdigit electrode pattern was used. The nanowires were dispersed in solution and drop cast onto the paper for the DEP to be performed, this was undertaken at

➤ To develop a simple proof of concept device – printable contacts were deposited onto filter paper

➤ Nanowire solution was drop cast and optimum DEP conditions were used align nanowires

➤ Ionic solutions consisting of Sodium Dodecyl Sulphate (SDS) diluted in DIW with varying concentrations were prepared, sensor paper was placed above solution

➤ The resultant current from the diode at a small negative bias was then observed

the optimum frequency previously determined. The DEP is controlled by the particles (nanowires) and medium (solvent) and as such the change in substrate should have a negligible effect. Once the solvent had evaporated and the device was ready the paper based device suspended directly above a salt based solution and gradually lowered till just in contact the solution. The paper based diode was biased small reverse voltage and the resultant current measured. This is a simple analogy to placing su sensor on the skin and utilizing the absorption of sw to measure the presence of particular biomarkers

25 Ionic sensor (I)

➤ To evaluate the performance we measured the current at a constant reverse bias of 10 mV, for solutions with varying concentrations of SDS

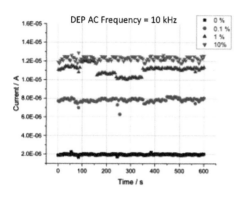

➤ A clear increase in current is observed as the SDS concentration increases

➤ There is an apparent plateau around 1%
  ➤ Paper Saturation?
  ➤ Nanowire Saturation?

clear increase in the current can be observed for increasing concentrations of SDS in the solution. appears to plateau at around 10% concentration, currently unclear if this is due to the paper

becoming saturated and unable to pass higher concentrations or due to the saturation of charge on the surface of the nanowires.

26 Ionic sensor (II)

➤To investigate the effect of utilizing only the "high quality" Nanowires we repeated this with a device realized from a sample fabricated with an AC frequency of 1000 Hz

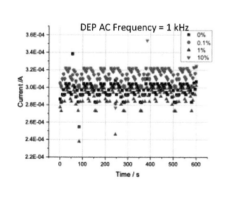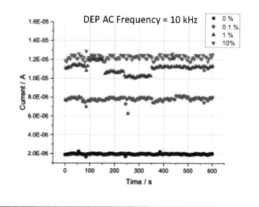

ilizing diodes fabricated at lower DEP frequencies hows a much weaker trend, indicating that the quality nanowires need to be selected via an

appropriate method if nanowires are to be used in a sensing system.

27 ## What we have achived

➢ Have demonstrated that DEP can be used to position InAs nanowires and realize electrical devices

➢ Have observed a relationship between AC frequency during DEP and the resultant NW conductivity and diode performance

➢ Demonstrated a proof of concept sensors for ionic liquids

➢ For InAs NWs have shown improved performance when only "high quality" NWs are selected during the fabrication

28 ## Remaining issues

• Paper based sensor needs to be integrated into a wearable system

  • At the moment nanowires just sense charge – need to target bio-markers
  • Packaging of sensor element (especially if its to be disposable / changeable)
  • Readout and control electronics – How will voltages be applied / currents measured?
  • Information display / readout – How will user receive information

Lots of work still remains, this can primarily be described in 4 areas: 1) Currently the nanowires are just detecting charge, to make a useful biosensor they will need to be functionalized so that only targeted biomarkers will attach to the nanowires, allowing sensitivity to select predetermined targets. 2) While this has demonstrated a paper based substrate can be used, this needs to be integrated packaged into a system in which the paper se element can be changed. 3) For use in a se additional electronics are needed to enable volta to be applied and the resultant currents recovered Finally for a wearable system, communication wi needed from the sensor to a suitable user interfa

29 Thanks

A. Marshall & A. Krier   (Nanowire growth at Lancaster)

S Laumier (PhD)
T. Farrow (PhD)
A Alsharmeri (PhD)
Z Cao (PhD)
N Sedgeti (post-doc)

Funding: EPSRC, DSTL, Royal Society

CHAPTER 08

Energy-Driven Computing for IoT

Domenico Balsamo

Newcastle University

Energy harvesting (EH) IoT systems are inherently accompanied by challenges in matching the dynamics of the power source with those of the power consumption. To mitigate this, systems typically incorporate large energy storage elements (referred to as 'energy-neutral' computing), increasing their volume, mass and cost. However, for future IoT systems that are highly constrained, it is instead desirable to minimize or even remove energy storage. In this presentation, I will discuss the emerging class of 'energy-driven' systems which reduce the energy storage by including the EH source and its availability as a primary consideration in the system's design. I will conclude my talk by exploring some open challenges necessitating further research to deliver on this vision of energy-driven IoT systems.

1 Outline

Research Vision
- IoT Nightmare: Power Availability
- Energy Harvesting
- From Energy-Neutral...
- ...to Energy-Driven Computing

Energy-Driven Computing
- Transient Computing
  - Hibernus and Hibernus++
- Power Neutral
  - Momentum
- Event-Aware Applications

...What about the communication?

2 IoT nightmare: Power availability

Ubiquitous computing dream of IoT systems everywhere is accompanied by the nightmare of battery replacement and disposal

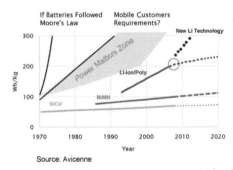

Source: Avicenne

Battery Technology is Stuck
No Moore's Law in batteries:
2-3%/year growth

IoT systems lifetime depends on battery life!

Solution

Design systems that harvest limited energy from ambient or scavenge power from human activity

Powering the Internet-of-things is a primary challenge for system designers as IoT systems typically have to last for many years, without intervention to charge or replace batteries.

Energy harvesting offers the potential for power systems to operate without batteries scavenging electrical power from environme sources or human activity.

3 Energy harvesting

**This term often refers to small autonomous devices –
micro energy harvesting (EH)**

Heat Motion and vibration  Light Ambient EM Energy

Thermal   Airflow   Photovoltaic Electromagnetic

Challenge

**Energy can be potentially infinite but the instantaneous power is
often uncontrollable as it depends on the source characteristics**

ifferent technologies of transducers can be used to convert different forms of energy into electricity.

4 From energy-neutral...

general, the wer from energy esting sources is cally characterized being spatially and porally dynamic and n uncontrollable. his is typically fied through the tion of a large gy storage to oth the temporal mics of supply and umption. his approach is vn as energy- ral operation.

Dynamic and uncontrollable *power generation*

- Energy-Neutral (EN) Operation:
  - EH systems include additional large energy storage to match:
    Energy consumed = Energy harvested
  - EH systems may also require the MPPT circuitry

Storage Drawbacks:
- Increased volume, weight and cost;
- Time to charge;
- Deteriorate in performance;

ever, adding energy storage reintroduces environmental and sustainability issues associated with batteries.

5   ... To energy driven (I)

- Research Questions:
  - Can systems be designed with this potential mismatch in mind, with execution being highly intertwined with power/energy availability?
  - Can systems be characterised by a relatively small (or even no) energy storage element compared to the average power consumption?

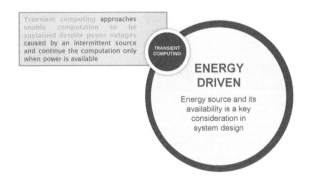

Transient computing approaches enable computation to be sustained despite power outages caused by an intermittent source and continue the computation only when power is available

TRANSIENT COMPUTING

ENERGY DRIVEN

Energy source and its availability is a key consideration in system design

6   Transient computing: Hibernus

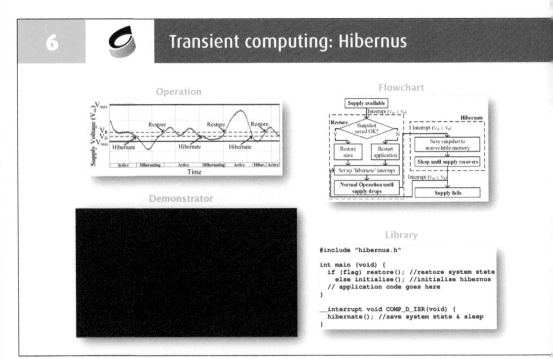

Operation

Flowchart

Demonstrator

Library

```
#include "hibernus.h"

int main (void) {
    if (flag) restore(); //restore system state
    else initialise(); //initialise hibernus
    // application code goes here
}

__interrupt void COMP_D_ISR(void) {
    hibernate(); //save system state & sleep
}
```

Hibernus enables computation to be sustained during intermittent supply. When the voltage drops below the system saves a snapshot and hibernates. When the voltage rises above Vr the system state is restored and continues operation.

7

Transient computing: Hibernus++

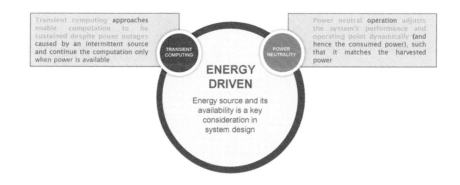

D. Balsamo *et al.*, "Hibernus++: A Self-Calibrating and Adaptive System for Transiently-Powered Embedded Devices," in *IEEE Transactions on Computer-Aided Design of Integrated Circuits and Systems*, vol. 35, no. 12, pp. 1968-1980, 2016.

Hibernus++ intelligently adapts the hibernate and restore thresholds in response to source dynamics and system load properties.

8

... To energy driven (II)

- **Research Questions:**
  - Can systems be designed with this potential mismatch in mind, with execution being highly intertwined with power/energy availability?
  - Can systems be characterised by a relatively small (or even no) energy storage element compared to the average power consumption?

Transient computing approaches enable computation to be sustained despite power outages caused by an intermittent source and continue the computation only when power is available

TRANSIENT COMPUTING

POWER NEUTRALITY

Power neutral operation adjusts the system's performance and operating point dynamically (and hence the consumed power), such that it matches the harvested power

ENERGY DRIVEN

Energy source and its availability is a key consideration in system design

9 — Power neutral: Momentum (I)

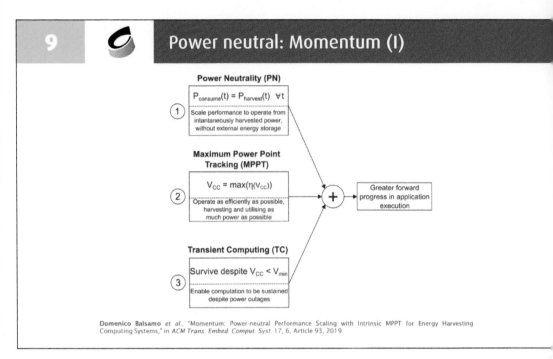

Power Neutrality (PN)

(1) $P_{consume}(t) = P_{harvest}(t) \ \forall t$

Scale performance to operate from intantaneously harvested power, without external energy storage

Maximum Power Point Tracking (MPPT)

(2) $V_{CC} = \max(\eta(V_{CC}))$

Operate as efficiently as possible, harvesting and utilising as much power as possible

Transient Computing (TC)

(3) Survive despite $V_{CC} < V_{min}$

Enable computation to be sustained despite power outages

Greater forward progress in application execution

Domenico Balsamo *et al.*, "Momentum: Power-neutral Performance Scaling with Intrinsic MPPT for Energy Harvesting Computing Systems," in *ACM Trans. Embed. Comput. Syst.* 17, 6, Article 93, 2019.

Momentum is a general power-neutral methodology, with intrinsic system-wide maximum power point tracking, that can be applied to a wide range of different computing systems, where the system dynamically scales performance (and hence power consumption) optimize computational progress depending on power availability.

10 — Power neutral: Momentum (II)

...based on dynamically adjusting the system's performance in real-time whilst tracking the harvested power...

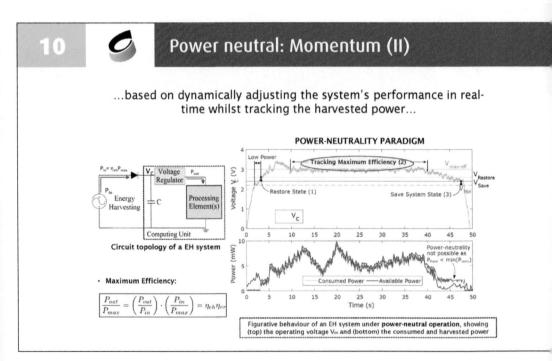

POWER-NEUTRALITY PARADIGM

$P_{in} = \eta_{eh} P_{max}$

Energy Harvesting — P_{in} — V_C Voltage Regulator — P_{out} — Processing Element(s)

Computing Unit

Circuit topology of a EH system

• **Maximum Efficiency:**

$$\frac{P_{out}}{P_{max}} = \left(\frac{P_{out}}{P_{in}}\right) \cdot \left(\frac{P_{in}}{P_{max}}\right) = \eta_{eh}\eta_{vr}$$

Low Power — Tracking Maximum Efficiency (2) — $V_{max\text{-}eff}$

Restore State (1) — Save System State (3) — $V_{Restore}$ / V_{Save}

V_C

Consumed Power —— Available Power — Power-neutrality not possible as $P_{harv} < \min(P_{cons})$

Figurative behaviour of an EH system under **power-neutral operation**, showing (top) the operating voltage V_{cc} and (bottom) the consumed and harvested power

We aim for the system to operate as close as possible to the maximum efficiency point. Here, the maximum efficiency point is defined as the combination of the EH source efficiency (as a fra of the maximum power point) and the efficien intermediate voltage regulation circuits.

11 ... Power neutral: Momentum (III)

...general power-neutral control strategy that enables the system to operate at its maximum efficiency...

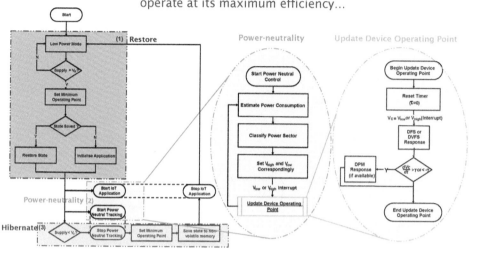

verall operation of Momentum.

12 ... To energy driven (III)

- **Research Questions:**
  - Can systems be designed with this potential mismatch in mind, with execution being highly intertwined with power/energy availability?
  - Can systems be characterised by a relatively small (or even no) energy storage element compared to the average power consumption?

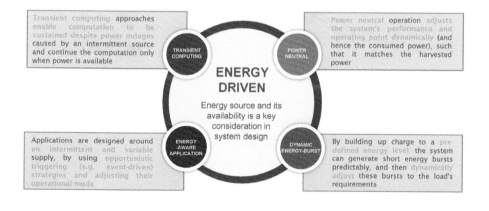

13   Energy-aware application

Traditional energy-neutral model

- Always-on
- Low-constant power in idle

Task-based model

- Transient-computing
  - Zero-power mode

Energy-aware model

- Keep track of time?
- Energy-driven QoS adjustment?
- Energy-driven approximate computing?

Moving from a traditional energy-neutral model to an energy-aware model, enables the zero-po" mode, using transient computing, minimising power consumption due to the idle mode.

14   ...What about the communication?

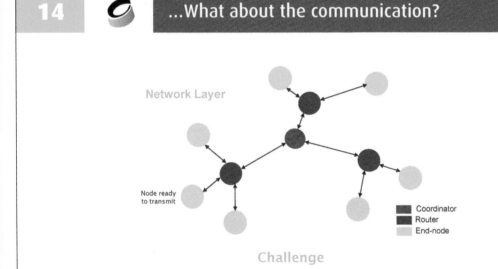

Network Layer

Node ready to transmit

■ Coordinator
■ Router
□ End-node

Challenge

The harvesting process ought to be incorporated into the network model to effectively utilize harvested energy

Another interesting and unsolved research challenge relates to the networking of intermittently-powe devices.

CHAPTER 09

Kinetic Energy Harvesters and Future Challenges for Iot and Wearable Devices

Elena Blokhina

University College Dublin

Dimitri Galayko

Sorbonne University

Kinetic energy harvesting is a technique to scavenge energy from the environment to power miniature autonomous sensors. In the first generation of energy harvesters, the designers relied on periodic motion to design and optimize the operation of a harvester. Since the functionality of sensors and the types of environment where they can be placed vary significantly, new techniques to scavenge kinetic energy from irregular motion, in particular the one produced by humans, have emerged. This talk overviews some key ideas of kinetic energy harvesting: availability of kinetic motion in a typical environment, mechanical and electrical components of an energy harvester, transduction mechanisms, system level requirements and future challenges in this field.

1 Content

- **Environment and energy harvesting**
  - Energy demands
  - Sources
- **Transducers and MEMS devices**
  - MEMS technology
  - Piezoelectric, electromagnetic and electrostatic transducers
  - State-of-the-art examples
- **Future challenges in kinetic energy harvesting**
  - System level considerations
  - Adaptive energy harvesting
  - Patterns in human motion

2   Environment and energy harvesting (I)

- Waves of interest towards energy harvesting[1]
- Driven by Wireless Sensors Networks (WSN) and the Internet of Things (IoT):
  - Each unit requires energy to operate
  - Batteries have limited lifetimes & must be replaced eventually
  - Replacement and recharge is difficult
  - Too many nodes predicted in the future
  - Location may be inaccessible
  - Sensor size and lifetime are limited by batteries
  - Let a sensor convert energy from the environment for its operation
- Energy harvesting (EH) must be taken very seriously after the most recent reports[2]

[1] P. Mitcheson, et al, "Architectures for vibration-driven micropower generators," JMEMS, vol. 13, 2004
[2] "Cisco visual networking index: Global mobile data traffic forecast update 2016–2021," Cisco, San Jose, CA, USA, White Paper, 2017

3 Environment and energy harvesting (II)

- New predictions:
  - 7 billion sensors in automotive industry now
  - 25 billion sensors connected to the Internet in 2 years
  - Power consumed by sensors is that of a small country
  - Demand for energy will outgrow any possible supply
- Energy "issue" stops the roll-out of the next wave of IoT
- Solution: energy harvesting in all possible forms
- Energy harvesting is the process of capturing and storing of energy
- Usually refers to small-scale autonomous devices and micro-scale energy harvesting

Mitcheson, et al, "Architectures for vibration-driven micropower generators," JMEMS, vol. 13, 2004
Cisco visual networking index: Global mobile data traffic forecast update 2016–2021," Cisco, San Jose, CA, USA, White Paper, 2017

4 Energy sources

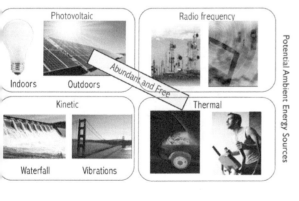

Estimated Harvested Energy		
Energy Source	Characteristics	Harvested Power
Light	Outdoor	100 mW/cm²
	Indoor	100 µW/cm²
Thermal	Human	60 µW/cm²
	Industrial	~1-10 mW/cm²
Vibration	~Hz–human ~kHz–machines	~4 µW/cm3 ~800 µW/cm3
Radio Frequency	GSM 900 MHz	0.1 µW/cm²
	WiFi	0.001 µW/cm²

Photovoltaic — Indoors — Outdoors

Radio frequency

Abundant and Free

Kinetic — Waterfall — Vibrations

Thermal

Potential Ambient Energy Sources

5 Applications of kinetic energy harvesting

♦ Medical and Health monitoring

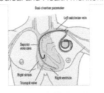

• Smart building

• Structure Health monitoring

Low data rate, low duty cycle, ultra-low power

• Wireless Sensor Networks

• Body Area Network

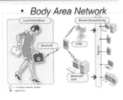

Deep Brain Neuro-stimulators

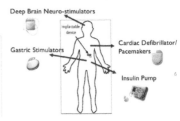

implantable device

Gastric Stimulators

Cardiac Defibrillator/ Pacemakers

Insulin Pump

---

6 Transducers and mems devices. Electromechanical transduction

- Transducer is a device that converts energy from one form to another

- For instance, "mechanical" signals (x, v or a) to "electrical" signals (i or v) or vice versa

- Energy harvesters are transducers

- Energy harvesters contain components (resonators) characterised by mechanical and electrical properties

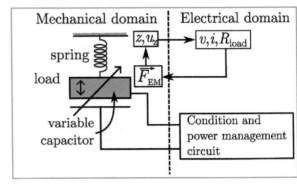

Mechanical domain | Electrical domain

spring

load

z, u_z → v, i, R_{load}

$\vec{F}_{\text{EM}}$

variable capacitor

Condition and power management circuit

7

Role of transducers in energy harvesting

- Vibrations available from the environment: low-frequency range; multiple frequencies

- Mechanism to

  - Capture one frequency through linear resonance response (2000 – 2010)

  - Capture multiple frequencies through wideband response (2010 – present)

  - Dynamical adjustment and frequency up-conversion (2015 – present)

- Realized through transducer design

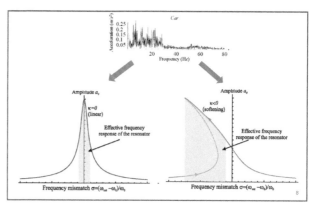

E. Blokhina et al. "Nonlinearity in Energy Harvesting Systems", Springer 2016

8

Mems transducers

- Micro-scale transducers are fabricated through MEMS technologies

- Similar to IC technology but allow one to make movable components

- Readily compatible with CMOS and integrated electronics

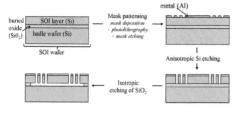

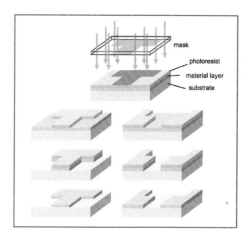

P. Basset, E. Blokhina, D. Galayko "Electrostatic Kinetic Energy Harvesting", ISTE Wiley, 2016

9 Mems resonant transducers

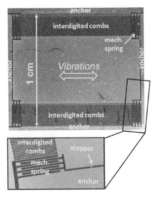

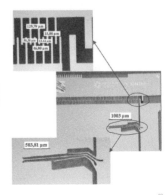

M. Dominguez, 2016; P. Basset, 2009   P. Basset, 2014-2016   E. Halvorsen, 2014-2016

---

10 Mems piezoelectric transducers (I)

- Employ the piezoelectric effect
- Displacement causes current and voltage
- Electromechanical coupling is considered linear ($V \propto \alpha\, x$) and weak
- Fragile devices
- Resonant operation
- Hybrid configurations reported

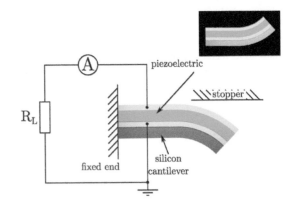

Oskar Z. Olszewski et al, Evaluation of Vibrational PiezoMEMS Harvester That Scavenges Energy From a Magnetic Field Surrounding an AC Current-Carrying Wire , JMEMS 2018

11 # Mems piezoelectric transducers (II)

- MEMS SOI
- 535 µm thick bulk silicon
- Piezoelectric AlN 500 nm
- 40 – 80 Hz resonant frequency
- 1.5 µW peak RMS power @ 2 A

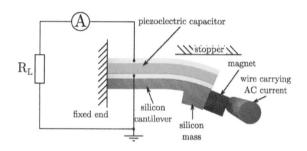

kar Z. Olszewski et al, *Evaluation of Vibrational PiezoMEMS Harvester That Scavenges Energy From a Magnetic Field Surrounding an AC rrent-Carrying Wire*, JMEMS 2018

---

12 # Mems piezoelectric transducers (III)

- MEMS SOI
- 535 µm thick bulk silicon
- Piezoelectric AlN 500 nm
- 40 – 80 Hz resonant frequency
- 1.5 µW peak RMS power @ 2 A

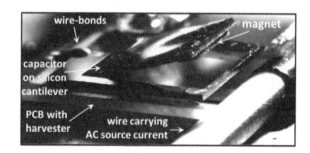

kar Z. Olszewski et al, *Evaluation of Vibrational PiezoMEMS Harvester That Scavenges Energy From a Magnetic Field Surrounding an AC rrent-Carrying Wire*, JMEMS 2018

13 Mems electromagnetic transducers (I)

- Employ Faraday's law
- Induced voltage is the rate of change of the magnetic flux:

$$\varepsilon = -\frac{d\Phi}{dt}$$

- Electromechanical coupling is considered linear ($V \propto \alpha\, v$) and weak
- Resonant and nonlinear resonant operation

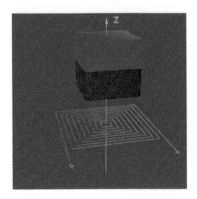

14

D. Mallick et al, "Surfing the high energy output branch of nonlinear energy harvesters," Phys Rev Lett, vol. 117, 2016.

14 Mems electromagnetic transducers (II)

- MEMS SOI
- $2.5 \times 2.5 \times 2.0\ mm^3$
- 300 – 600 Hz resonant frequency
- Few µW peak power @ 1g

15

D. Mallick et al, "Surfing the high energy output branch of nonlinear energy harvesters," Phys Rev Lett, vol. 117, 2016.

15 Mems electrostatic transducers (I)

- Employ the principle of variable capacitors:

$$C(t) = f(x(t))$$

- Displacement causes capacitance variation $C(t)$ and hence variable voltage and current in the transducer

- Electromechanical coupling is nonlinear and strong

- Resonant, nonlinear resonant and controlled operation

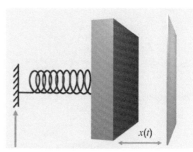

External acceleration acts on the frame and causes one plate to move

Plate moves and distance can change

$x(t)$

16

16 Mems electrostatic transducers (II)

- Many possibilities to improve the design of transducers to achieve a noticeable variation between C_{max} and C_{min} and create a wideband response:

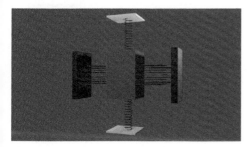

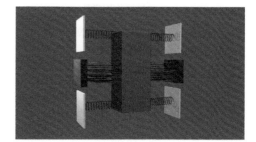

17

17 Mems electrostatic transducers (III)

Comb-drive, gap closing motion [Basset, JMM, 2013]

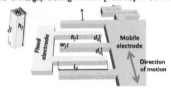

Multilayer, comb geometry, area overlapping motion
[Deterre et al., J. of Physics, 2013]

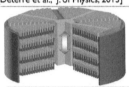

Plane, gap closing motion [Miao et al., Microsyst.
Technologies, 2006]

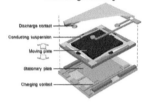

Comb geometry, area overlapping motion
[Le et al., J. Intel. Mat. Syst. Struct., 2012]

18 Mems transducers comparison

Type	Advantages and Disadvantages	
Piezoelectric (pz)	**Advantages:** - No need for biasing or pre-charging **Disadvantages:** - Fragile - Weak electromechanical coupling	
Electromagnetic (em)	**Advantages:** - No need for biasing or pre-charging **Disadvantages:** - Difficult to integrate magnetic components - Weak electromechanical coupling	
Electrostatic (es)	**Advantages:** - Strong coupling **Disadvantages:** - Nonlinear devices - Need for biasing and pre-charging	

E. Blokhina et al. "Nonlinearity in Energy Harvesting Systems", Springer 2016

19 Future challenges in kinetic energy harvesting. system level considerations

$$P_{max} = 4\, m\, A_{ext} \cdot f_{ext}\, X_{lim}$$

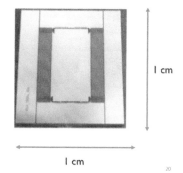

- Maximum power that can be extracted from external vibrations with 1 cm³ volume device and sinusoidal external vibrations:
  - X_{lim} = 50 µm
  - $m = 0.5 \times 1\,cm^3 \times \rho_{si}$ = 1.2 gram
  - A_{ext} = 1g and f_{ext} = 100 Hz
  - $P_{max} \approx 240$ µW
- This figure is optimistic, with an optimal transduction mechanism
- Never reached in practice because of technological limitations
- State-of-the-art results: 10 – 20 µW

1 cm

1 cm

20

---

20 Examples

- MEMS SOI
- Larger bandwidth (1~180 Hz)
- The energy per cycle (maximum) 450 nJ/cycle
- Bias voltage 45 V
- Acceleration 11 Hz and 3 g_{peak}
- Powered 3 RFID communications within 16 s

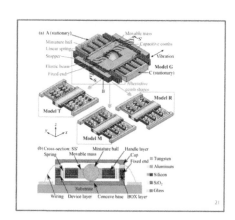

Lu et al, "A power supply module for autonomous portable electronics: ultralow-frequency MEMS ectrostatic kinetic energy harvester with a comb structure reducing air damping", Microsystems & anoengineering, 2018

21

21 Future directions: non-resonant energy harvesting (I

- Most of reported energy harvesters are resonant
  - Their operation is optimized for harmonic inputs
  - Limited range of input frequencies
- Explore non-resonant solutions
  - No hypothesis is made on the form of the input external vibrations
- How to synthetize the transducer force for both control and energy conversion to maximize the converted energy?
  - "synthetize the force" = smart controls needed
  - Control electronics consumption can be made low
  - Enough that the gain is much greater than that of resonant solutions
- This is a change in the dominant paradigm of energy harvesting

P. Basset, E. Blokhina, D. Galayko "Electrostatic Kinetic Energy Harvesting", ISTE Wiley, 2016

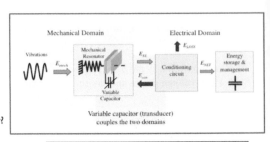

Variable capacitor (transducer) couples the two domains

Electrical and mechanical domains are coupled in all realisations of kinetic energy harvesters

Conditioning electronics can be used to extract energy from the mechanical domain and to control the mobile mass

---

22 How much energy is in external vibrations?

- Arbitrary waveform is characterised by a set of extremal values $a_{max}(t)$ and $a_{min}(t)$

- The maximum energy W_{lim} that can be extracted from an arbitrary waveform:

$$W_{lim} = 2X_{lim}m\sum(a_{max,i} - a_{min,i})$$

- The converted power is

$$P_{lim} = 2X_{lim}m\sum(a_{max,i} - a_{min,i}) \cdot \frac{1}{t_{max} - t_{min}}$$

E. Blokhina, et al, "Pattern Recognition in Human Motion for Kinetic Energy Harvesting", ICECS 2019, Genova, Italy.
D. Galayko. et al, "Energy Harvesting for the IoT: Perspectives and Challenges for the Next Decade", ICECS 2018, Bordeaux, France

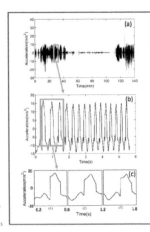

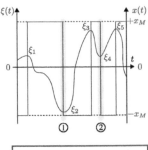

Need to know in advance when to toggle the mass

We invest energy but gain more as a result

23 Future directions: Non-resonant energy harvesting (II)

- "Toggling" of the proof mass costs energy

- "Toggling" energy cost = fixed cost + proportional cost

- The toggling may not be profitable at a given maximum/minimum

- Since we invest energy in order to gain more, we must invest it in the toggling of the proof mass at the right moment when the combination of maximum/minimum is optimal

- **Smart decision block is required**

- **Smart conditioning electronics is required**

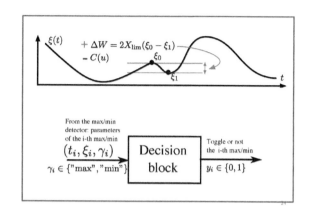

D. Galayko, et al "AI Opportunities for Increased Energy Autonomy of Low Power IoT Devices" ICECS 2019, Genova, Italy.

24 Is prediction possible for human motion? (I)

- Broad range of time series analysis techniques

- Methods originating from the areas of

  - Random processes

  - Dynamical systems

  - Machine learning

- Simple illustration of patterns in some type of human motion using the cosine-similarity index:

$$\text{Similarity} = \frac{\mathbf{A} \cdot \mathbf{B}}{|\mathbf{A}| \cdot |\mathbf{B}|} = \frac{\sum_{i=1}^{N} A_i B_i}{\sqrt{\sum_{i=1}^{N} A_i^2} \cdot \sqrt{\sum_{i=1}^{N} B_i^2}} \in [-1, 1]$$

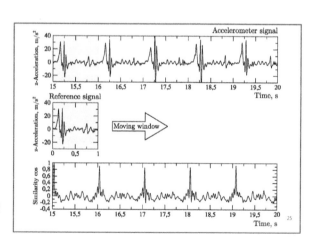

25 Is prediction possible for human motion? (II)

- Walking or running can display recognizable patterns
- However no universal motion patterns across general day-to-day activities
- Not possible to train the decision block once and use for any kind of activity
- Algorithms of the decision block and their circuit implementations are feasible
- Future of energy harvesting – complex systems and adaptive approach to process available energy in the environment
- Ongoing work: implementation of the decision block [new results will be shown during the presentation]
- Ongoing work: implementation of control circuit

26

26 Conclusions

- The area is active
- First generation of EH offered a fundamental idea but it was not efficient
- EHs are advanced now, inherently complex and nonlinear
- New prototypes operate in strongly nonlinear mode and are able to supply a sensor
- Harvesters need complex conditioning and power management electronics
- Harvesters need to be adaptive to operate with irregular vibration patterns
- Future energy harvesting systems will be complex
- No serious study of reliability
- No serious study of costs, effectiveness and operation in a real environment

27

About
the Editors

ara Ghoreishizadeh

ara Ghoreishizadeh is a lecturer (assistant professor) in wearable technologies at Aspire Centre for Rehabilitation and Assistive technology (Aspire CREATe) University College London (UCL). She is also a member of the UCL Institute Healthcare Engineering (IHE). She received a PhD from Ecole Polytechnique lerale de Lausanne (EPFL), Switzerland in 2015 and has been with the Dept. Electrical and Electronic Engineering, Imperial College, as a postdoctoral earch Fellow during 2015 to 2018. Her research interests include analogue/ ced-signal IC design for biomedical applications, monolithic integration of sensors with microelectronics, and innovative microchip based technologies assessment of acute pain. Her research has been funded by Wellcome Trust, RC, Imperial College London, and UCL Biomedical Research Council (BRC).

2 has authored/co-authored over 35 publications in international journals or ference proceedings. She acts as a reviewer for several journals and conferences is on the editorial board of the Journal of Microelectronics.

Ghoreishizadeh is an elected member of the IEEE BioCAS Technical Committee the Early Career Researcher (ECR) steering committee of the EPSRC eFutures work. She serves on the organisation committee of several IEEE conferences uding ISCAS, BioCAS, ICECS, MeMea, and NewCAS, as Technical chair, Track ir, Technical Programme Committee member, Review Committee member, session chair. She co-initiated and co-organise(d) the 1st and 2nd UK Circuits Systems (UKCAS) Workshops in 2018 and 2019. She is a member of IEEE, IET, uits and Systems (CAS), Solid-State Circuit (SSC), and Engineering in Medicine Biology (EMB) societies.

Kylie de Jager

Kylie de Jager is a Postdoctoral Research Associate in the Asp Centre for Rehabilitation and Assistive Technology (CREATe), University College London (UCL). She completed a B.Eng degree Electrical & Electronic Engineering, at the Rand Afrikaans Univer (RAU) and a M.Eng degree in Power Electronics at the University Johannesburg (UJ) before receiving a PhD from University Colle London in Medical Physics and Bioengineering. She spent time a Visiting Scholar within the Centre for Power Electronics Syste (CPES) at Virginia Tech and worked as a Postdoctoral Research Fell in the Division of Biomedical Engineering (BME) at the University Cape Town (UCT) with whom she still holds an Honorary Resea Affiliate position. She is also a member of the IEEE.

Her research interests focus on the design and development of med electronics which has included: implantable electronics; neuromuscu stimulators; biosignal amplifiers; and, stimulation artefact reducti She also conducts research within the field of health innovat to investigate medical device development paradigms to be understand the context in which healthcare needs are address She has co-authored over 30 peer-reviewed journal and confere publications.

About
the Authors

Domenico Balsamo

Domenico Balsamo received his B.Sc. and M.Sc. degrees Electronic Engineering from the University of Modena Reggio Emilia, Italy, in 2005 and 2008, respectively. He then recei his Ph.D. degree in Computer Engineering from the University Bologna, Italy, in 2015. He was appointed as a Lecturer in Embedd Systems Design and IoT at the Newcastle University in 2019. was previously employed as a research fellow at the Univer of Southampton. His research broadly covers energy-effici embedded systems and low-power pervasive computing, on wh he published over 30 scientific papers in highly respected jourr and top-notch conference proceedings.

ena Blokhina

lena Blokhina received the Habilitation HDR (equiv. D.Sc.) degree in electronic engineering from UPMC Sorbonne Universities, France, in 2017, the Ph.D. gree in physical and mathematical sciences and the M.Sc degree in physics n Saratov State University, Russia, in 2006 and 2002 respectively. From 2005 2007 she was a research scientist at Saratov State University. Since 2007, she been with the School of Electrical and Electronic Engineering of University lege Dublin, Ireland, and is currently an associate professor and Head of the uits and Systems Research Group.

f Blokhina is a Senior member of IEEE and the Chair of the IEEE Technical nmittee on Nonlinear Circuits and Systems. She had been elected to serve a member of the Boards of Governors of the IEEE Circuits and Systems Society S) for the term 2013-2015 and has been re-elected for the term 2015-2017. n 2014 to 2017, she chaired the IEEE CAS Young Professionals and Women circuits and Systems committee. She has served as a member of organising nmittees, review and programme committee, session chair and track chair at ny international conferences on circuits and systems and nonlinear dynamics uding as IEEE International Symposium on Circuits and Systems (ISCAS), IEEE n American Symposium on Circuits and Systems (LASCAS), IEEE Asia-Pacific ference on Circuits and Systems (APCCAS), International Symposium on linear Theory and Its Applications (NOLTA), IEEE international Conference on tronics, Circuits and Systems (ICECS) and others. Prof Blokhina has served as Programme Chair of the first edition of IEEE Next Generation of Circuits and tems Conference (NGCAS) 2017 and of IEEE ICECS 2018. In 2016-2017 Prof hina was an Associate Editor for IEEE Transactions on Circuits and Systems I, since 2018 she is the Deputy Editor in Chief of that Journal.

research interests include the analysis, design, modelling and simulations of linear circuits, systems and networks with particular application to complex, erging and mixed-domain multi-physics systems.

Alex Casson

Dr Alex Casson is a Associate Professor (Reader) in the Materia Devices and Systems division in the Department of Electri and Electronic Engineering at the University of Manchester. Casson's research focuses on human body sensing in out-of-the-l artefact rich, power constrained situations. Typical applications are brainwave monitoring, transcranial stimulation for neuromodulati and cardiac monitoring. He also has extensive expertise in low pov sensor nodes with onboard real-time signal processing, particula wearable sensors for human monitoring where signal processing used to decrease power consumption for energy harvester powe systems.

Dr Casson gained his undergraduate degree from the University Oxford in 2006 where he read Engineering Science specialising Electronic Engineering (MEng). He completed his PhD from Impe College London in 2010, winning the prize for best doctoral thesis electrical and electronic engineering. Dr Casson worked as a resea associate and research fellow at Imperial College until 2013 when joined the faculty at the University of Manchester. He is currentl Senior Member of the IEEE, Chair of the Institution of Engineering a Technology's healthcare technologies network, and lead the Univers of Manchester's bioelectronics activities.

imitri Galayko

imitri Galayko received his PhD degree in 2002 from the University Lille-I in 2002. Since 2005 he is an Associate Professor Sorbonne Université (formerly UPMC) in LIP6. The design and delling of systems for vibration energy harvesting has been one is main research topics since 2007, on which he has collaborated the Circuits and Systems Group at UCD, Ireland, and with the rosystem group of the University of Paris-Est, France. In 2016, co-authored 2 books in the topic of energy harvesting systems, lished by Springer and Wiley-ISTE. He was a coordinator of two ich national collaborative research grants (ANR), HODISS (ARFU gram, 2009-2012) et HERODOTOS (Arpege program, 2011-2014), he was local scientific leader of the ANR grant SESAM. His lication record is 28 articles in international journals, 90 articles ternation-al conferences, and 4 patents. He was associate editor he IEEE Trans. On Circuits and Systems II journal in 2016-2017. He esently a local coordinator for Sorbonne University of ERASMUS+ ect APPLE. He supervised 10 PhD dissertations.

Pantelis Georgiou

Pantelis Georgiou currently holds the position of Reader (Associate Professor) in Biomed Electronics at Imperial College London within the Department of Electrical and Electrc Engineering. He is the head of the Bio-inspired Metabolic Technology Laboratory in Centre for Bio-Inspired Technology; a multi-disciplinary group that invents, develops a demonstrates advanced micro-devices to meet global challenges in biomedical scie and healthcare. His research includes ultra-low power micro-electronics, bio-inspired circ and systems, lab-on-chip technology and application of micro-electronic technology create novel medical devices. Application areas include new technologies for treatm of diabetes such as the artificial pancreas, novel Lab-on-Chip technology for genomics a diagnostics targeted towards infectious disease and antimicrobial resistance (AMR), a wearable technologies for rehabilitation of chronic conditions.

Dr. Georgiou graduated with a 1st Class Honours MEng Degree in Electrical and Electronic Enginee in 2004 and Ph.D. degree in 2008 both from Imperial College London. He then joined the Institut Biomedical Engineering as Research Associate until 2010, when he was appointed Head of the I inspired Metabolic Technology Laboratory. In 2011, he joined the Department of Electrical & Electr Engineering, where he currently holds an academic faculty position. He conducted pioneering work the silicon beta cell and is now leading the project forward to the development of the first bio-insp artificial pancreas for treatment of Type I diabetes. In addition to this, he made significant contribut to the development of integrated chemical-sensing systems in CMOS for Lab-on-Chip applicat used in rapid diagnostics for infectious diseases. Dr. Georgiou is a senior member of the IEEE and and serves on the BioCAS and Sensory Systems technical committees of the IEEE CAS Society. He i associate editor of the IEEE Sensors and TBioCAS journals. He is also the CAS representative on the sensors council. In 2013 he was awarded the IET Mike Sergeant Achievement Medal for his outstand contributions to engineering and leading a multidisciplinary team to deliver innovative medical devi In 2017, he was also awarded the IEEE Sensors Technical Achievement award in the area of Se systems for significant contributions in bioelectronics, and in 2018 he was awarded the Rosetrees Interdisciplinary Award for the development of a rapid diagnostic test to accurately detect bact infection in children using microchip technology. He is also an IEEE Distinguished Lecturer in Circuits Systems, a role focused on shaping the global community in his field.

adi Heidari

adi Heidari (PhD, SMIEEE, FHEA) is a Lecturer in the James Watt School of Engineering at the University of Glasgow, UK. He is ding the Microelectronics Laboratory (www.melabresearch. n) and his research includes developing magnetoelectronics for urotechnology devices. He has authored over 100 publications in -tier journals and conferences.

Heidari is a member of the IEEE Circuits and Systems Society Board Governors (2018-2020), IEEE Sensors Council Member-at-Large 20-2021), Senior Member of IEEE and Fellow of Higher Education idemy (FHEA). He is an Associate Editor for the IEEE Journal of ctromagnetics, RF and Microwaves in Medicine and Biology and E Access, Editor of Elsevier Microelectronics Journal, and Guest or for the IEEE Sensors Journal, and Frontiers in Neuroscience. is the General Chair of 27th IEEE ICECS 2020, Technical Program ir of IEEE PRIME'19, and serves on the organising committee of eral conferences including the UK Circuits and Systems Workshop CAS), UK-China Emerging Technologies (UCET) Conference, IEEE SORS'16 and '17, NGCAS'17, BioCAS'18, PRIME'15, ISCAS›23, and organiser of several special sessions on the IEEE Conferences. His earch has been funded by major research councils and funding anizations including the European Commission, EPSRC, Royal iety and Scottish Funding Council. He is part of the €8.4M EU H2020 Proactive project on "Hybrid Enhanced Regenerative Medicine tems (HERMES)".

has been the recipient of a number of awards including the 2019 Sensors Council Young Professional Award, Rewards for Excellence e from University of Glasgow (2018), IEEE CASS Scholarship CAS'17), Best Paper Awards from the ISSCC'16, ISCAS'14, and ME'14 and '19.

Elizabeth Rendon-Morales

D r Rendon-Morales is a Senior lecturer in Electrical and Electron Engineering in the Department of Engineering and Design at t University of Sussex. The main area of her research is the desi development and testing of the next generation of sensing electro systems and medical instrumentation. Her areas of expertise inclu sensors, electronics, robotics and telemetry systems. Within sensing area, She is leading the development of advanced sens devices to monitor electrocardiogram -ECG signals on babies dur early pregnancy and throughout labour. On the robotics area, She leading the development of micron level precision instrumentat tools to achieve linearity and repeatability that could contribute the next generation surgical autonomous robotic systems.

n Sandall

n currently a lecturer in the Department of Electrical Engineering
id Electronics at the University of Liverpool, Prior to this I obtained
PhD in Physics from Cardiff University in 2007 and have previously
ked for Philips Research in Eindhoven, and at The University of
ffield. My research primarily concerns the development and
lication of semiconductor based electronic and photonic sensors.

current research is focused on utilizing semiconductor materials
devices, including nanowires, quantum dots and thin films for
mical and biological sensing applications. These include point-of-
, lab-on-a-chip and wearable approaches for in-vivo, in-vitro and
itro diagnosis and monitoring of a range of health conditions.

Stephen Taylor

Stephen Taylor is currently Reader in Musculoskele Instrumentation and Telemetry at the Institute of Orthopaed and Musculoskeletal Science (IOMS), UCL. He has developed a implanted in man femoral and shoulder instrumented orthopae implants, for wireless measurement of forces during activity. He conducted clinical trials to study implant loosening and isome implant forces. He designed a novel magnetic drive for the extendible prosthesis for paediatric limb lengthening. His curr research interests are: in vivo telemetry of measured forces fr within spinal rods, shoulder and elbow joints and tendon for musculoskeletal model validation; wheelchair instrumentation accessibility and injury prevention. He is author or co-author of Journal papers.

Ian Underwood

Ian Underwood is an award winning academic, innovator and serial technology entrepreneur (https://www.research.ed.ac.uk/portal/en/persons/ian-underwood.html).

He recently stepped down after ten years as Head of the Research Institute for Integrated Micro and Nano Systems (IMNS) and remains Deputy Director of the EPSRC Centre for Doctoral Training in Intelligent Sensing and Measurement (CDT-ISM). Current research interests include photonics, electronic displays, liquid crystal and OLED microdisplays, micro- and nano-scale sensors, sensor arrays and applications of miniature sensors ranging biomedical and healthcare to automotive and space.

As a PhD student at the University of Edinburgh in the late 1980s he was a pioneer of the new technology of liquid crystal on silicon, developing it first for spatial light modulators and later for microdisplays. He spent a formative year as a Fulbright fellow at The University of Colorado in the early 1990s where he helped design the first product for new spin-out Boulder Nonlinear Systems. Returning to the University of Edinburgh he helped found Micropix Technologies in 1995 (now Forth Dimension Displays) then left the university to co-found MicroEmissive Displays in 1998. MicroEmissive developed the smallest colour tv-screen in the world (Guinness Book of World Records 2004). More recently he has been an advisor to start-ups Holoxica and Optoscribe and founding chairman of Pure VLC (now Pure LiFi). He is currently supporting commercialisation of SmartScroll (https://www.eng.ed.ac.uk/research/projects/smartscroll).

He has been accepted into fellowship of the Royal Academy of Engineering, where he is a mentor at the Enterprise Hub; the Royal Society of Edinburgh where he chairs the selection panel

for enterprise fellowships; and the Institute of Physics. At the Socie
for Information Display he is Publications Chair, with oversight of t
journal, magazine and book series.

Underwood is co-author of the SID/Wiley book entitled Introduction
Microdisplays (Wiley, 2006) and author of six chapters of the Handbc
of Visual Display Technology (Springer 2011 and 2016). From 20
to 2012 he served on the Scottish Science Advisory Council that off
impartial advice to the Scottish Government on matters of scier
policy. He was lead author on the 2014 innovation report entit
Making the Most of our Scientific Excellence.

Printed and bound by CPI Group (UK) Ltd, Croydon, CR0 4YY

23/10/2024

01777693-0017